DR. JOE SCHWARCZ

Published by ECW Press
665 Gerrard Street East
Toronto, Ontario, Canada M4M 1Y2
416-694-3348 / info@ecwpress.com

Editor: Jen Albert
Cover design: David A. Gee

LIBRARY AND ARCHIVES CANADA CATALOGUING IN PUBLICATION

Title: Better not burn your toast : the science of food and health / Dr. Joe Schwarcz.

Names: Schwarcz, Joe, author

Description: Includes index.

Identifiers: Canadiana (print) 20250217287 | Canadiana (ebook) 20250217325

ISBN 978-1-77041-791-5 (softcover)
ISBN 978-1-77852-474-5 (ePub)
ISBN 978-1-77852-475-2 (PDF)

Subjects: LCSH: Nutrition—Popular works. | LCSH: Food science—Popular works.

Classification: LCC QP141 .S35 2025 | DDC 612.3—dc23

This book is funded in part by the Government of Canada. *Ce livre est financé en partie par le gouvernement du Canada.* We also acknowledge the support of the Government of Ontario through the Ontario Book Publishing Tax Credit, and through Ontario Creates.

Canada

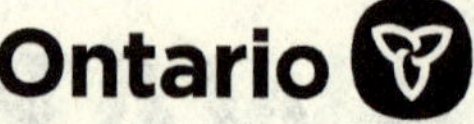

PRINTED AND BOUND IN CANADA

PRINTING: FRIESENS 5 4 3 2 1

ECW Press is a proudly independent, Canadian-owned book publisher. Find out how we make our books better at ecwpress.com/about-our-books

The interior of this book is printed on Sustana EnviroBook™, which is made from 100% recycled fibres and processed chlorine-free.

PRAISE FOR *SUPERFOODS, SILKWORMS, AND SPANDEX*

"This highly informative, authoritative title makes solid science accessible and entertaining, and it keeps alive the author's tradition of clearly differentiating pseudoscience and quackery from empirical science. Schwarcz's fans will love this latest book, and he'll likely gain a new following as well."
— *Library Journal*, starred review

PRAISE FOR *QUACK QUACK*

"In his usual inimitable fashion, Joe Schwarcz weaves tales of pseudoscience, showing how modern quackery is very similar in many ways to quackery from decades and even hundreds of years ago. The details change, but the core concepts behind quackery tend to remain disturbingly the same, as do the sales techniques. It's an informative and entertaining read for anyone interested in recognizing quackery when they see it."
— David H. Gorski, professor of surgery, Wayne State University School of Medicine

PRAISE FOR *SCIENCE GOES VIRAL*

"Joe Schwarcz does it again with this fun, fast-paced, and evidence-informed exploration of the hot topics in science we've been bombarded with over the past few years! From the biology of vaccines (including the new mRNA variety) and immune boosting (spoiler: you can't) to the history of epidemiology and toilet paper, Schwarcz gives us the fact-filled low-down. In a world filled with misinformation and twisted science, this is a must-read!"
— Timothy Caulfield, Canada Research Chair in Health Law and Policy, bestselling author of *Is Gwyneth Paltrow Wrong about Everything?*

PRAISE FOR *A GRAIN OF SALT*

"Schwarcz's light touches of humor make the scientific information feel accessible and ensure that it's entertaining. With enough facts to soothe anxious, health-conscious individuals as well as some good tidbits to share, this enlightening collection offers every reader something new to learn and marvel over."
— *Booklist*

PRAISE FOR *A FEAST OF SCIENCE*

"Huzzah! Dr. Joe does it again! Another masterwork of demarcating non-science from science and more generally nonsense from sense. The world needs his discernment."
— Dr. Brian Alters, professor, Chapman University

PRAISE FOR *THE FLY IN THE OINTMENT*

"Joe Schwarcz has done it again. In fact, he has outdone it. This book is every bit as entertaining, informative, and authoritative as his previous celebrated collections, but contains enriched social fiber and 10 percent more attitude per chapter. Whether he's assessing the legacy of Rachel Carson, coping with penile underachievement in alligators, or revealing the curdling secrets of cheese, Schwarcz never fails to fascinate."
— Curt Supplee, former science editor, *Washington Post*

PRAISE FOR *DR. JOE & WHAT YOU DIDN'T KNOW*

"Any science writer can come up with the answers. But only Dr. Joe can turn the world's most fascinating questions into a compelling journey through the great scientific mysteries of everyday life. *Dr. Joe and What You Didn't Know* proves yet again that all great science springs from the curiosity of asking the simple question . . . and that Dr. Joe is one of the great science storytellers with both all the questions and answers."
— Paul Lewis, president and general manager, Discovery Channel

PRAISE FOR *THAT'S THE WAY THE COOKIE CRUMBLES*

"Schwarcz explains science in such a calm, compelling manner, you can't help but heed his words. How else to explain why I'm now stir-frying cabbage for dinner and seeing its cruciferous cousins — broccoli, cauliflower, and brussels sprouts — in a delicious new light?"
— Cynthia David, *Toronto Star*

PRAISE FOR *RADAR, HULA HOOPS AND PLAYFUL PIGS*

"It is hard to believe that anyone could be drawn to such a dull and smelly subject as chemistry — until, that is, one picks up Joe Schwarcz's book and is reminded that with every breath and feeling one is experiencing chemistry. Falling in love, we all know, is a matter of the right chemistry. Schwarcz gets his chemistry right and hooks his readers."
— John C. Polanyi, Nobel Laureate in Chemistry

Also by Dr. Joe Schwarcz

Superfoods, Silkworms, and Spandex: Science and Pseudoscience in Everyday Life

Quack Quack: The Threat of Pseudoscience

Science Goes Viral: Captivating Accounts of Science in Everyday Life

A Grain of Salt: The Science and Pseudoscience of What We Eat

A Feast of Science: Intriguing Morsels from the Science of Everyday Life

Monkeys, Myths, and Molecules: Separating Fact from Fiction, and the Science of Everyday Life

Is That a Fact?: Frauds, Quacks, and the Real Science of Everyday Life

The Right Chemistry: 108 Enlightening, Nutritious, Health-Conscious and Occasionally Bizarre Inquiries into the Science of Everyday Life

Dr. Joe's Health Lab: 164 Amazing Insights into the Science of Medicine, Nutrition and Well-Being

Dr. Joe's Brain Sparks: 179 Inspiring and Enlightening Inquiries into the Science of Everyday Life

Dr. Joe's Science, Sense and Nonsense: 61 Nourishing, Healthy, Bunk-Free Commentaries on the Chemistry That Affects Us All

Brain Fuel: 199 Mind-Expanding Inquiries into the Science of Everyday Life

An Apple a Day: The Myths, Misconceptions and Truths About the Foods We Eat

Let Them Eat Flax!: 70 All-New Commentaries on the Science of Everyday Food & Life

The Fly in the Ointment: 70 Fascinating Commentaries on the Science of Everyday Life

Dr. Joe & What You Didn't Know: 177 Fascinating Questions &Answers About the Chemistry of Everyday Life

That's the Way the Cookie Crumbles: 62 All-New Commentaries on the Fascinating Chemistry of Everyday Life

The Genie in the Bottle: 64 All-New Commentaries on the Fascinating Chemistry of Everyday Life

Radar, Hula Hoops and Playful Pigs: 67 Digestible Commentaries on the Fascinating Chemistry of Everyday Life

TABLE OF CONTENTS

SCIENCE AND UNCERTAINTY

Our word "science" derives from the Latin for "knowledge." But knowledge is actually quite a complicated concept. Sometimes we can make a categorical, conclusive statement about having knowledge, but much more frequently, claims of knowledge have to be qualified with "ifs," "buts," and "maybes," terms that describe a degree of uncertainty. Indeed, certainty in science can be elusive.

If I were asked if I know what will happen if vinegar is added to baking soda, I can answer with total confidence that bubbles will form. But how do I really know that? Because I have done this many times, as have millions of others since 1846 when baking soda was first introduced, and bubbles have always formed. There has never been a single report of this not happening. And of course, based on chemistry, I know that reacting acetic acid with sodium bicarbonate produces carbon dioxide gas. So, I am certain of my answer.

Now, what if I were asked if I know what will happen if a feather and a hammer are simultaneously dropped from the same height on the moon? My answer would be that they hit the ground at the same time. How confident am I of this answer? Very. Like anyone who has studied physics, I learned how Galileo proved wrong Aristotle's theory about falling objects. The famous Greek philosopher had stated that if two objects of different mass were dropped from the same height, the heavier one would hit the ground first. It seemed logical, but Aristotle was not an experimentalist.

Some twenty centuries later, Galileo Galilei was. Whether he actually performed the experiment of simultaneously dropping cannon balls of different mass from the leaning tower of Pisa is debatable, but he did roll balls down a ramp at different angles and extrapolated his observations to a sheer vertical drop, concluding that the time taken for a dropped object to hit the ground was independent of its mass. Why then did a feather take a longer time to fall than a heavier object? Air resistance! This was proven in 1971 when astronaut David

Scott, commander of Apollo 15, dropped a hammer and a feather on the moon from the same height and, as we clearly saw on television, they both hit the surface at the same time. Putting aside the ludicrous conspiracy theory that the moon landing and the video were faked, we can say that we know, without a doubt, that Galileo was right. There is no need to ask that this experiment be repeated by other astronauts to confirm the results.

Let's now turn to another question I have been repeatedly asked recently. Does the presence of perfluoroalkyl substances (PFAS) in our drinking water pose a risk to health? I would have to say, maybe, but I don't really know. These chemicals, due to their oil-resistant and water-resistant properties, are found in numerous consumer items ranging from stain-free carpets and rain gear to cookware and cosmetics. Since they are manufactured on an immense scale globally, it comes as no surprise that traces can be detected in our drinking water, especially given that today's sophisticated analytical techniques can detect substances down to concentrations of parts per trillion.

Laboratory experiments with cell cultures and animals have shown that PFAS can have an effect on health and that these chemicals can be detected in virtually everyone's blood. What we do not know is whether the amounts detected in blood are clinically significant, or to what extent their presence comes from water. In order to have a conclusive answer, one would have to compare blood levels and disease patterns in a group of subjects who drank tap water with a control group who only drank water guaranteed to be free of these chemicals. Such a study, which would have to span several years, is logistically and economically not feasible, so we are left with the "maybe" conclusion.

Such uncertainty does not only apply to substances in our drinking water. There is uncertainty when it comes to the effects of COVID vaccines, the best diets for weight control, the effects of the microbiome on health, genetically modified foods, vitamin supplements, sleeping aids, coffee, saturated fats, red wine consumption, antioxidants, and a whole host of other factors that can impact our health and longevity.

However, even uncertainty is on a sliding scale and is subject to qualifications. For example, population studies have shown that people with cardiovascular risk factors such as high cholesterol, a family history of heart disease, or who are overweight are likely to benefit from taking statins. But whether a given individual will avoid a heart attack or stroke cannot be predicted with confidence. Similarly, while COVID vaccines reduce the risk of infection, not everyone who is vaccinated benefits. Decisions come down to a risk-benefit evaluation, and even that is subject to change as more studies come to light. While limited alcohol consumption used to be regarded as innocuous, or even beneficial, recent studies suggest that alcohol is a carcinogen and it may be that no amount of alcohol is safe.

This uncertainty, which is inherent to many aspects of science, can drive one batty. One study finds that breakfast is the most important meal of the day, another argues that it is better to eliminate it. One study shows that artificial sweeteners are a great tool for weight loss, while another claims that not only are they useless, but unhealthy. One study highlights the benefits of canola oil, another portrays it as deadly. On and on it goes.

But let's not despair. Absolute certainty in science is indeed rare, but as studies accumulate and point in the same direction, we can arrive at a pretty high degree of certainty. For example, I'm quite confident in saying that cutting back on sugar is a good idea, as is incorporating berries and whole grains into the diet. Exercise is an important determinant of health, vaccines do not cause autism, smoking is bad, crystals have no healing power, spoons cannot be bent by the power of the mind, and homeopathic preparations are no more effective than placebos. And I'll also go on record stating that I know, without any doubt, that the Earth is not flat, even though I have not had the experience of seeing its spherical shape from space. Some things we know, some we are uncertain of. Science has many but not all answers. This is especially true when it comes to matters of health. There is much we know and much of which we are uncertain. Let's explore!

TO WORRY, OR NOT TO WORRY. THAT IS THE QUESTION.

"Three Thousand Six Hundred Food Packaging Chemicals Detected in Human Bodies: How Reheating Food Is Killing Us." That was the headline that greeted me last week from one of the news services to which I subscribe. There were many others in the same vein. I knew that a slew of emails from worried people would follow wanting to know what to make of this. After all, nobody wants their obituary to state "Killed by reheating food."

The actual title of the paper published in the *Journal of Exposure Science & Environmental Epidemiology* was "Evidence for Widespread Human Exposure to Food Contact Chemicals." It is no secret that our food and beverages meet up with numerous "food contact materials" before they contact our lips. During production, they can pass through plastic or metal pipes and encounter various types of processing machinery, ranging from slicers and dicers to conveyor belts. They are then packaged in glass, paper, plastic, or metal containers that feature an assortment of inks and adhesives. Food contact materials themselves have to be produced, and that involves a very large number of chemicals. When it comes to plastics, the different varieties are made from different chemicals and use different plasticizers, stabilizers, catalysts, and preservatives. There are remnants of the monomers used to make the polymers as well as various polymer degradation products. Paper production involves about two hundred different chemicals such as pigments, bleaching agents, and coatings.

Around 14,000 chemicals have been found to be present in food contact materials. That this can be determined is a testimonial to the talents of analytical chemists and the manufacturers of their instruments. There is no question that some of these chemicals can migrate into foods and beverages; however, it must be remembered that the presence of a chemical cannot be equated to the presence of risk!

A number of these food contact chemicals have been studied in terms of their potential toxicity. For example, aluminum, bisphenol A, phthalates, and perfluoroalkyl substances (PFAS) have been investigated for possible carcinogenic, and endocrine- and metabolism-disrupting effects with some disturbing results. But these studies have mostly used cell cultures and animals. Human epidemiological studies have also associated some of these chemicals with medical conditions, but of course association is not the same as causation.

Knowing that 14,000 chemicals may be present in food contact materials is one thing, but it is the number that end up in our body that is of interest. That's the question that the authors of the paper that generated the headlines tackled. They scrutinized five biomonitoring programs that had analyzed blood and urine samples from thousands of people for a range of chemicals and also examined worldwide databases that have compiled data from a variety of studies investigating human exposure to potential toxins. In all, some 3,600 food contact chemicals were identified as having been detected in human bodies and about 150 of these are judged to be of concern based on cell culture and animal data.

What are we to make of this? I just don't know. We are exposed on a regular basis to thousands and thousands of chemicals from personal care products, cleaning agents, air pollutants, medications, food, and water. A classic example is coffee with over a thousand compounds present, including carcinogens such as acrylamide, furfural, and benzopyrene, yet coffee does not cause cancer. Even with carcinogens, dose matters.

Besides these food contact chemicals invading our bodies, there are so many other things to worry about. Arsenic in rice? How about phthalates? They are in our shower curtain, nail polish, and our children's PVC duckies. How about sodium lauryl sulphate in shampoos? Mercury in our dental fillings? Or in fish? Mycotoxins in cereal? *E. coli* in meat? Listeria in cold cuts? Salmonella in eggs? Parabens in cosmetics? Or siloxanes? Lead in lipstick? Antimony in our bottled water? Gluten in wheat? Estrogens in soy? Hormones in milk? Polycyclic hydrocarbons in steak? Acrylamide in potato chips? Pesticide residues on fruit? Formaldehyde outgassing from wrinkle-free shirts? Or from particle board kitchen cabinets?

If that isn't enough, we can worry about aluminum in antiperspirants and perchloroethylene residues in dry-cleaned clothes. Then there are the flame retardants from our couch that accost us as we watch television which floods us with a variety of programs about all the toxins in our life. We can also worry about artificial sweeteners and flavors. And let's not forget food dyes or monosodium glutamate. Paradichlorobenzene in urinal cakes? Hexane in our cooking oil? Volatile organic chemicals in paint? Antibiotics in meat? Bisphenol A in cash register receipts? Or in white dental fillings?

There's more. Titanium dioxide nanoparticles in sun protection products. UV exposure if we don't use sun protection products. High-fructose corn syrup. Diesel fumes. Diacetyl in popcorn flavoring. Aflatoxins on peanuts. Cobalt and chromium leaching from medical implants. PCBs in window caulking. Caramel in colas. Paraphenylenediamine in hair dyes. Chlorine in baby carrots. Petrolatum in skin, lip, and hair products. Artificial musk in fragrances. Butylated hydroxyl toluene in makeup. Prescription drugs in tap water. Carcinogens oozing from crumb rubber in artificial turf. Nonoxynol in detergents. Brominated vegetable oil in beverages. Polycyclic aromatic hydrocarbons in driveway sealants. Methylene chloride in paint stripper. Acrylonitrile in synthetic fabrics. Dioxane in bubble bath. If all of this drives you to drink, well reconsider, because ethanol is a known carcinogen!

What's the point? There are scientific papers published about all these concerns, often prompting alarmist headlines. The fact is that we are exposed to a vast array of potential toxins in various combinations, and it is essentially impossible to know what effects they have in the doses to which we are exposed. Of course, efforts should be made to reduce exposure to substances like bisphenol A, phthalates, and pesticides that have a toxic cloud hanging over their heads. But this needs to be addressed at the manufacturing level. Consumers can drive themselves crazy by trying to avoid "toxins," and the associated stress is certainly detrimental to health.

As far as "reheating food killing us" goes, there is no mention of any such thing in the paper about food contact chemicals. The headline writer probably remembered something about chemicals leeching out from plastics in the microwave. Indeed, it is good advice to use glass or ceramic in the microwave, but suggesting that "reheating food" precipitates an appointment with the Grim Reaper is alarmist nonsense.

CALCIUM PROPIONATE IN OUR DAILY BREAD? NO WORRIES!

"Now you know why he carries his baguette in this fashion!" Chuckles usually erupt when I make this comment in a talk on food additives as I simultaneously show a picture of a man sporting a beret and holding a bottle of wine, making him instantly identifiable as French. The kicker is that he has a baguette tucked under his arm. My little joke caps a discussion of the use of calcium propionate as a preservative in bread.

Of course, there is a backstory here. One that goes way, way back. Archaeologists have found what look like grinding stones dating back some 30,000 years, possibly for grinding wild grains into food. That's conjecture, but finding 14,000-year-old charred crumbs of flatbread in the area that is present-day Jordan is believed to be proof of early attempts at bread making. It's likely that wild grains were ground and mixed with water to make a paste that was then poured onto hot stones.

Sometime later, it was noted that if this paste were put aside for a few days, it would rise. Dough was born! Baking this in primitive ovens gave rise to bread. Of course, at the time there was no knowledge that it was carbon dioxide generated by naturally occurring yeasts that make the dough rise, and that yeast also produces a host of compounds giving bread a delectable flavor. This stimulated the cultivation of wheat beginning some 10,000 years ago, and we have been enjoying bread ever since.

However, the road to that enjoyment has been marked by a number of potholes. Most recently, there has been concern about gluten, a network of proteins found in wheat that traps carbon dioxide and is responsible for the texture of bread. Some unfortunate people, generally estimated to be less than 1 percent of the population, are afflicted with celiac disease, a chronic, serious condition that is triggered by the ingestion of gluten and affects the digestive and immune systems. There is no question that celiacs have to scrupulously eliminate all traces of gluten from their diet. But many "wellness influencers" with no background in science also urge all their followers to avoid gluten claiming that it generally impairs health. While there is evidence of some consumers of gluten experiencing short term belly ache and bloating, a condition termed "non-celiac gluten sensitivity," it is rare, afflicting less than 6 percent of the population. Nevertheless, this has given rise to a marketing bonanza of gluten-free products, demonstrating the ease with which pseudoscience slithers into our lives.

A pothole that was encountered long before the gluten issue was spoilage. Naturally occurring fat in grains can react with oxygen in the air and break down to produce compounds that can give bread a "rancid" taste and smell. There is no health danger here, but molds are a different story. These microbes, present in soil and air, can get a foothold in the moist, nutrient-rich environment of bread and multiply. This is more than a cosmetic concern because molds can produce toxic metabolites. And then there is a problem that has plagued bakers since ancient times, a condition known as *ropiness*. This is well named because it is characterized by the inside of a loaf being converted into a mass that resembles a tangle of strings that make for an unpleasant chewiness. To make matters worse, rope formation is accompanied by the release of volatile compounds such as diacetyl, acetoin, acetaldehyde, and isovaleraldehyde that produce a scent resembling that of rotting melons or pineapples.

Ropiness is caused by the action of several bacterial species of the genus *Bacillus*. These do not cause disease, but they release enzymes that decompose wheat starch and proteins into smelly compounds. At the same time, they secrete their own proteins and polysaccharides that cause discoloration and slimy rope formation. These bacteria and their dormant forms called spores are ubiquitous and are found in plants, soils, and even baker's yeast. While heat destroys the bacteria, their spores survive baking. The moist and low-acid conditions present in bread make for an ideal environment for the spores to burst into life and cause mischief.

By the early twentieth century, the science of bacteriology, spurred by Louis Pasteur's pioneering studies of microbes, had been established. Since it was clear that bacteria play a role in disease and food spoilage, the search for antibacterial substances was on! Acids were found to be the enemies of bacteria, so these served as a natural starting point. An interesting clue was discovered in Emmental cheese, produced since the thirteenth century in Switzerland. The holes in the cheese are the result of carbon dioxide being formed by bacteria that are naturally present in raw milk. These also form propionic acid that imparts a characteristic flavor to the cheese and, given the long history of the cheese, do not pose a health concern. Various experiments with propionic acid and its salts were undertaken with the conclusion that calcium propionate added to bread dough prevents contamination by molds and interferes with the growth of bacteria that produce ropiness.

Calcium propionate is now commonly added to bakery flour to an extent of 0.3 percent, which is less than the 1 percent found naturally in Emmental cheese. Sourdough is an exception because its content of lactic acid produced by the added bacterial culture serves the same purpose as propionates. The safety of calcium propionate was underlined by the discovery that propionates are a normal product of fat metabolism and that we naturally produce these compounds in our body. Today, calcium propionate is produced by neutralizing propionic acid with calcium hydroxide, with the propionic acid in turn being made by reacting ethylene with carbon monoxide and water using a cobalt or rhodium catalyst.

Now back to my Frenchman. The French are famous for the attention they give to food, particularly their pastries and baguettes. Indeed, a law passed in 1993 regulates that any bread sold as "traditional" cannot contain additives. Only flour, water, salt, baker's yeast, and lactic acid producing cultures are allowed. The absence of calcium propionate is not a problem because traditional baguettes are consumed the same day they are baked, so there is no time for *Bacillus* spores to awaken and cause ropiness. And the ropiness is further prevented by carrying the baguette under the arm. Since we naturally produce propionates in our gut, from where they enter our bloodstream, and thence our sweat, the traditional French method of transport makes for an effective natural preservative. HaHaHa.

CONCERNS ABOUT SMOKED FOODS HEAT UP

Cavemen in the Stone Age didn't have many luxuries. They had, however, learned to make fire by banging together flintstones or rubbing two sticks together and that at least made their cold and dingy dwellings more comfortable. And fire did something else as well. Once the caveman had clubbed his prey, which of course wasn't a dinosaur, fire allowed the meat to be cooked. If the meat wasn't eaten right away it would be hung to dry because experience had shown that dried meat keeps longer, especially if exposed to smoke. Much later, science would demonstrate that bacteria require moisture to proliferate, hence the preservative action of drying, and that smoke contributes to the effect due to the antibacterial compounds it contains. There was another benefit. The meat tasted better! And we have been enjoying smoked meat ever since. At least until studies revealed that where there is smoke, there may be more than one kind of fire.

That fire is in the form of some nasty compounds in smoke, with the term *carcinogen* rearing its ugly head. Smoke contains literally hundreds of compounds, a number of which, particularly polycyclic aromatic hydrocarbons (PAHs) and heterocyclic amines (HCAs) are of concern. In cell culture and animal studies these have been shown to cause changes in DNA that can lead to cancer. But cells and animals aren't people. However, some epidemiological studies have also linked the consumption of smoked meats to intestinal cancer. For example, in one region of Hungary where home-smoked foods are a major part of the diet, the incidence of stomach cancer is twice that in the rest of the country. Analysis found meats here to contain an average of 4.16 micrograms per kilogram of benzopyrene, one of the polycyclic aromatic hydrocarbons. This is roughly six times the average of 0.7 micrograms per kilogram detected in commercially produced smoked foods in the rest of the country. The European Union sets a safety limit of 2 micrograms per kilogram of benzopyrene in meat.

Besides the carcinogens already present in smoke, the heat associated with smoking can also result in chemical reactions in the meat that form yet more carcinogens such as acrylamide, acrolein, and nitrosamines. The latter are formed as the result of nitrogen and oxygen in the air forming nitric oxide that then reacts with naturally occurring amines in the meat. Unfortunately, these concerns also come into play with barbecued foods that owe at least some of their delightful taste to smoke.

It turns out that food does not actually have to be smoked in a smokehouse or grilled on a barbecue to have the flavor of smoke. That's thanks to an accidental discovery made in 1895 by Missouri pharmacist Ernest Wright. As the story goes, Wright had noted a drop of liquid trickling down a stove pipe and surmised that it was the result of smoke condensing on the cooler surface. Could this be liquefied smoke, he wondered? A taste confirmed that it was indeed "liquid smoke." Maybe food could be smoked, he thought, without the need for a smokehouse.

Wright designed an apparatus, essentially a still, in which smoke from burning hickory wood ran through a cooling column where it condenses into a liquid. He made a fortune by selling his liquid smoke not as a flavoring, but as a preservative. Competitors, seeking another angle, entered the business and promoted liquid smoke as a flavoring agent. The marketplace was soon flooded with smoked cheese, oysters, potato chips, soups, hot dogs, and even bacon that never saw a smokehouse. Then the European Food Safety Authority (EFSA) threw a wrench into the works in 2013 by questioning the safety of liquid smoke. No regulations were introduced at the time, but it was recommended that liquid smoke be used in reduced quantities.

Ten years later, EFSA changed its tune and concluded that liquid smoke flavoring cannot be considered safe for human health. What changed? Technology! More sensitive tests had been developed for genotoxicity, the property of a chemical to damage a cell's genetic information. Also, analytical techniques to identify the various components of liquid smoke had improved. For example, 2-furanone and pyrocatechol now joined the other genotoxic compounds found in smoke. Not only were these present, but their amounts exceeded the "threshold of toxicological concern (TTC)." For chemicals that can damage DNA the TTC is defined as a daily intake of 0.0025 micrograms per kilogram of body weight, roughly 1.5 micrograms for an adult. EFSA now concluded that "no safe concentrations can be defined for such substances." However, EFSA is not a regulatory agency, so European countries will have to decide whether the use and sale of liquid smoke should be curbed.

Canadian and American regulatory agencies so far do not see any risk with liquid smoke as it is used. Actually, there are fewer carcinogens in liquid smoke than in traditionally smoked meat. That's because when smoke condenses, it forms an aqueous layer and a heavier oily layer. The oily layer contains the fat-soluble compounds like the polycyclic aromatic hydrocarbons, while most of the flavorful compounds are in the water layer which is the one used for liquid smoke. Furthermore, compounds such as 2-furanone are found in other foods that have been heat treated as well. For example, one study detected higher than TTC levels of 2-furanone in chocolate sampled from chocolate cakes and crepes.

To sum up, liquid smoke, which is always used in small amounts, is hardly a health concern. In smoked potato chips or smoked cheese, for example, the fat content is a bigger issue than the trace carcinogens in the smoke flavoring. By contrast, foods that are exposed to real smoke will contain far more carcinogens. A serving of smoked herring, smoked salmon, or, I wish I didn't have to say this, Montreal-style smoked meat, will all deliver far more than the daily threshold of 0.15 micrograms of genotoxic chemicals. Moral of the story? Don't eat these foods every day! What about asking if Montreal-style smoked meat could be made with liquid smoke instead of a smoker? Don't even let that question roll off your tongue!

BETTER NOT BURN YOUR TOAST

I will sheepishly tell you that I set off the fire alarm in my office when I was preparing my morning ritual of avocado toast and didn't notice that one of the slices touched an element in the toaster until I smelled smoke. It seems the smoke detector is very sensitive! Luckily it was early morning and only a few people had to evacuate the building. Then I had a quandary. Eat the toast or chuck it? That question was raised because I was familiar with the scientific literature that had assessed the risks of eating burned foods, particularly as it pertains to acrylamide, a purported carcinogen that forms when foods containing both carbohydrates and the amino acid asparagine are heated to temperatures that exceed about 120 degrees Celsius. Bread, essentially made of starch with small amounts of asparagine, falls into that category.

When heated, some of the starch breaks down to release glucose, and proteins decompose to release amino acids to join the amino acids already present. Glucose and amino acids can then engage in what is known as the *Maillard reaction*, named after French chemist Louis Camille Maillard who in 1912 described the reaction between sugars and amino acids that results in an array of compounds that give browned foods their distinctive flavor. When that amino acid is asparagine, the compound that forms is acrylamide, classified by the International Agency for Research on Cancer (IARC) as "a probable human carcinogen." This is based on feeding large doses to animals and making an educated guess about the effect of smaller doses on humans.

Different science agencies make different guesses, varying from 25 to 195 micrograms, about the maximum amount that an adult can safely ingest every day. These are based on animal data because, obviously, humans cannot be fed different amounts of acrylamide and be monitored for decades to determine the incidence of cancer. The closest one can come are studies that follow the health status of groups of people who periodically fill out food frequency questionnaires from which acrylamide intake can be estimated. The majority of such studies have found no association with cancer.

Nevertheless, it is prudent to try to minimize exposure to any substance that causes cancer in animals, so we can look at acrylamide content of specific foods and compare it to the guesses for maximum recommended daily intake. When calculations are made taking all foods into account, an adult consumes a daily average of 30 to 40 micrograms, well below the average of the guesses. Potato chips and french fries are at the high end, with a serving containing about 50 micrograms. A serving of cereal has about 7 micrograms and a cup of coffee less than 1. And toast? That's roughly 5 micrograms per slice. Burned toast would have more but still below the most stringent daily recommended intake of 25 micrograms.

Although the connection of acrylamide in humans is tenuous, researchers have investigated various methods to reduce exposure. An obvious goal is the reduction of asparagine that is present in foods that also contain sugars and free amino acids. If there is no asparagine, acrylamide cannot form. Asparaginase is an enzyme that catalyzes the conversion of asparagine to unreactive aspartic acid and can be isolated from a variety of fungi and bacteria. The common mold *Aspergillus niger* and a strain of *E. coli* are typical examples. Asparaginase added to flour should then be able to reduce the amount of acrylamide that forms when dough made from this flour is heated.

Italian researchers explored this possibility by looking at, what else, pizza. The amount of acrylamide in the final product was reduced by 50 percent! Other scientists demonstrated a 90 percent reduction of acrylamide in toast made from dough treated with asparaginase and a close to 60 percent reduction in french fries from potatoes that had been soaked in an asparaginase solution.

Another attractive approach is to reduce the amount of asparagine that is naturally present in wheat and potatoes through genetic engineering. The production of asparagine in grains and potatoes requires the activity of several genes, suggesting that if these genes are silenced to some degree, the amount of asparagine produced is reduced. Two possibilities arise. Either prevent the genes from giving the signal that triggers the production of asparagine or remove these genes from the wheat or potato's DNA.

The first can be accomplished by RNA silencing. The message to produce asparagine involves transferring the information needed for its formation from DNA in the cell's nucleus to messenger RNA (mRNA) that then uses this information to direct the cell to make asparagine. In 2006, Andrew Fire and Craig Mello were awarded the Nobel Prize in Physiology or Medicine for discovering that short strands of RNA that can be synthesized in the lab can bind to and inactivate a specific mRNA. This technology has already been used to develop potatoes that reduce the potential formation of acrylamide. Since no foreign genes are introduced, there is no labeling requirement. Interestingly french fry marketers have not jumped to promote their use of such potatoes because it would suggest that the fries they were selling before had an element of risk.

For wheat, the technology that holds promise is based on the tool CRISPR-Cas9, which garnered the Nobel Prize in Chemistry for Emmanuelle Charpentier and Jennifer Doudna in 2020. It is usually described as a type of molecular scissors, allowing for specific genes, such as the ones responsible for producing asparagine, to be edited out from DNA. Again, no foreign genes are introduced. Professor Nigel Halford at the Rothamsted Research center in the U.K. has used the CRISPR-Cas9 technique to develop wheat that has a greatly reduced asparagine content. The research has emerged from the laboratory into field trials that have demonstrated that the technology works.

There is no question that the methods to reduce our intake of acrylamide through these technologies are of significant academic interest. However, their practical significance is questionable, given that the evidence of acrylamide being a human carcinogen is less than compelling. The average Canadian adult consumes about 25 micrograms of acrylamide a day, an amount that even by the European Food Safety Association's and California Proposition 65's strict standards is not a problem.

Now back to my burned toast. Knowing that I wouldn't be consuming anything further that day with any significant acrylamide content such as chips or french fries, I scraped off the black stuff, spread the slice with avocado, and ate it. But I am now more careful not to burn my toast. And more importantly, to limit my consumption of chips and french fries.

THE FASCINATING STORY BEHIND THE DISCOVERY OF FREE RADICALS

You wouldn't think that a paper published in 1900 in the *Journal of the American Chemical Society* with the rather unimposing title "An Instance of Trivalent Carbon: Triphenylmethyl" would have a huge impact on life. But did it ever! That paper described the first ever production of a free radical, a highly reactive species destined to become a major player in our understanding of numerous aspects of life, ranging from diseases and their treatment to the production of plastics. Free radicals would even offer a glimpse into the process of aging and possible dietary modifications to slow it down.

The story, however, starts with a different kind of radical, one that ended a life. The life was that of Russia's Tsar Alexander II, the radical was Ignacy Hryniewiecki who hoped that assassinating Alexander would lead to an overthrow of the tsarist autocracy. The bomb he hurled at Alexander killed the Tsar but not the tsarist regime. The 1881 assassination triggered unrest in Russia and a series of pogroms against Jews even though the assassin was a Catholic. Alexander's successor, Tsar Alexander III in 1882 passed the May Laws, which imposed residency and business restrictions on Jews. As a result, Hirsh Gomberg's farm was confiscated and in 1884 the family, including eighteen-year-old son Moses, fled to the United States as did some two million other Russian Jews.

In Chicago, while attending high school, young Moses worked under brutal conditions in the stockyards that would be described in Upton Sinclair's epic 1905 book *The Jungle*. Moses learned English and in 1886 was accepted into the University of Michigan where he went on to obtain a doctorate in chemistry based on his research into the chemistry of caffeine. Since at the time Germany was the hotbed of chemical research, Gomberg decided that his career could be best furthered by working with one of the German luminaries. Adolf von Baeyer was one such, already having achieved fame for his synthesis of indigo. Working under von Baeyer, Gomberg became interested in making molecules that were notoriously difficult to synthesize. One of these was hexaphenylethane in which two carbon atoms are bonded together with each one attached to three bulky benzene rings. The thinking was that the large benzene rings could not be made to fit happily around the two carbons.

Gomberg's idea was to start with triphenylchloromethane, a compound that had three benzene rings and a chlorine atom attached to a central carbon. He hoped that removing the chlorine atom would allow the denuded carbon atoms to join together. There was precedent for this because French chemist Charles Wurtz had shown that such coupling reactions were possible when molecules containing chlorine attached to carbon were treated with sodium.

When Gomberg tried this reaction, he did get a product, but its elemental composition, that is the percent of carbon and hydrogen, did not match that of the desired hexamethylethane. It did match a structure in which three benzene rings are attached to a single carbon. But this seemed impossible! German chemist August Kekule, who had been von Baeyer's mentor, and Scottish chemist Archibald Scott Couper had in 1858 simultaneously established that the carbon atom always forms four bonds. Now Gomberg was proposing that he had actually isolated a substance in which carbon was "trivalent" that is, it formed only three bonds.

Upon his return to the U.S. in 1899, he was appointed assistant professor at the University of Michigan where he would spend the rest of his career, eventually serving as chair of the chemistry department. Just a year after his appointment, Moses Gomberg published his classic paper in which he proposed that he had managed to produce a molecule in which carbon formed only three bonds, and he suggested that such a species would be highly reactive since carbon was not happy to be in this deprived state. He described a number of reactions that proved that this *free radical*, as the novel species was termed, was indeed extremely reactive and formed products that were predictable based upon its trivalent structure. Gomberg recognized that this discovery could have far-reaching consequences and ended his paper with the rather curious statement, "This work will be continued and I wish to reserve the field to myself."

That was not to happen! Thousands and thousands of papers have been published since then dealing with free radicals although at first the chemical community was very skeptical of Gomberg's claim. It took thirty years before the existence of free radicals was accepted with a critical step being taken by University of Chicago chemist Morris

Kharasch, another Russian Jewish immigrant, who demonstrated that some reactions that seemed to be inexplicable could in fact be explained if free radicals were involved. He then went on to show that highly reactive free radicals could initiate a chain reaction leading to the formation of polymers. This would prove to be of critical importance during the Second World War with the production of synthetic rubber using free radical polymerization.

However, these days we are more likely to connect free radicals to biological effects and to diets that feature antioxidants, substances that neutralize them. That takes us back to 1954 when University of Rochester biologist Rebeca Gerschman proposed that during the process of cellular respiration, as glucose reacts with oxygen to produce energy, reactive oxygen species (ROS), including free radicals, are created and cause cells to age and die. This too was received with skepticism and was not proven to be correct until 1969 when Irwin Fridovich at Duke University discovered an enzyme that was christened *superoxide dismutase* because it was found to catalyze the destruction of a free radical, termed *superoxide*, that was a product of respiration as Gerschman had suggested.

Denham Harman had become intrigued by Gerschman's 1954 paper because he had been interested in the aging process since 1945 when his wife had shown him a magazine article about Russian scientist Alexander Bogomolets's efforts to prolong life. At that time, Harman was working as a research chemist for Shell Oil, but after being captivated by the article he decided to go to medical school because he believed that would provide a background for the anti-aging research in which he was interested. On learning of Gerschman's oxygen free radical theory, he recalled his days at Shell when he had been working on free radical polymerization and was familiar with antioxidants that interfered with the process. In 1956, Dr. Harman published a landmark paper in the *Journal of Gerontology* in which he proposed his free radical theory of aging. He then went on to show that dietary antioxidants administered to mice countered the effects of free radicals and prolonged their lives. Despite this never having been demonstrated in humans, a giant industry of antioxidant supplements has mushroomed based on unsubstantiated claims. I think Moses Gomberg would not be amused.

JELLYFISH AND MUSHROOMS

Sherlock Holmes, that is to say, Sir Arthur Conan Doyle, didn't get it quite right. The sting of *Cyanea capillata*, commonly known as the lion's mane jellyfish, is not as "dangerous to life as the bite of the cobra." Yet that is just how Holmes describes the potency of the jellyfish's sting in "The Adventure of the Lion's Mane." It is his belief about the toxicity of the venom that leads him to conclude that what at first seemed to be a murder was actually death due to an encounter with *Cyanea capillata*. As a medical doctor, Conan Doyle should have known that while the tentacles of the creature can deliver a very painful sting, they are not a murder weapon. To give the author a bit of leeway, it turns out that the victim in the story had a pre-existing heart condition that could have contributed to his death.

Cyanea capillata, which really does resemble the mane of a lion with its 1,200 or so tentacles, each of which can be several meters long, is an impressive creature. The longest tentacles ever recorded were thirty-six meters, making this jellyfish the largest animal in the world in terms of length. A slap on the skin by a tentacle results in a long red welt due to the venom, a complex mixture of histamine, kinins, prostaglandins, and tryptamines. While *Cyanea* can be let off the hook as a killer, other jellyfish can be indicted for such crimes. The most famous one is *Chironex fleckeri*, commonly known as the sea wasp, which can inject a neurotoxic venom consisting of over sixty proteins that can kill an unfortunate swimmer within a few minutes.

In "The Adventure of the Lion's Mane," Holmes explains that he knew about the lion's mane because he is an "omnivorous reader with a strangely retentive memory for trifles." Based on that description of himself, one can guess that the world's most famous consulting detective would have been familiar with another version of the lion's mane, the mushroom that goes by that name. Like the jellyfish, *Hericium erinaceus* bears a resemblance to the mane of a lion. It is an edible mushroom that can be found growing on the trunks of trees with a history of use in traditional Chinese medicine, where it is said to counter symptoms of a deficiency in qi, the vital life force that flows through the body. By restoring the flow of qi, the mushroom supposedly promotes good digestion, general vigor, and strength. Recently lion's mane has become trendy, generally in the form of some sort of extract, with claims of various health benefits including anticancer activity, better brain function, regeneration of nerves, improved immunity, reduced anxiety, effective control of blood sugar, and lowering of cholesterol. That is quite a litany of claims.

What would Sherlock Holmes say, if asked about the possibility of lion's mane having such a diversity of medicinal effects? A good bet is that the detective, who habitually insisted on theories being based on facts instead of twisting facts to conform to theories, would have simply asked, "Where is the evidence?" In Victorian times, there would not have been any. Today, Holmes would be directed towards a number of research papers that have investigated the potential of the mushroom as a therapeutic agent.

Mushrooms, along with yeasts and molds, are fungi, living organisms that are neither plants nor animals since they do not photosynthesize and must therefore obtain all the substances they need for life by absorption from their environment. Yeasts have a long history of use in the production of bread, wine, and beer. As far as the medicinal properties of mushrooms are concerned, their use has been largely been anecdotal, save for the use of "magic mushrooms" or fly agaric (*Amanita muscaria*), both capable of inducing hallucinations. Lewis Carroll's 1865 *Alice in Wonderland* has the heroine consuming a mushroom, "one side of which makes her taller and the other side makes her shorter." We can only guess at what mushroom Carroll had in mind, but the effects of *Amanita muscaria* were well known to the Victorians, and the red mushrooms with white dots were a common presence on Victorian Christmas cards.

The first medical use of fungi, and still the most famous one, was the production of penicillin by a mold, famously discovered by Alexander Fleming in 1928. The success of penicillin as an antibiotic generated much interest in all sorts of fungi and led to the mushrooming of the scientific literature devoted to the possible health benefits of these organisms. Lion's mane in particular captured the attention of researchers because of its purported multiple benefits. Various extracts have been shown to act as antibiotics, inhibit cancer cell growth, reduce inflammation, and regrow damaged nerve cells. While intriguing, these studies were carried out in cell cultures or in animals, but as we well know, the human body is not a giant test tube and neither are we seventy kilogram rodents.

The greatest potential for lion's mane, or more specifically, certain of its components, seems to lie in protecting the nervous system from damage and in improving cognition. Benefits have been seen in animal models of Parkinson's and Alzheimer's disease, and for brain function, there are a couple of those rare entities, a human trial. In a double-blind, placebo-controlled study, a lion's mane supplement was found to improve cognitive function in a group of senior Japanese subjects with mild cognitive impairment. In another clinical trial, daily consumption of cookies containing two grams of the mushroom resulted in a reduction of some symptoms of anxiety and depression in menopausal women.

All of this is interesting, but it is difficult to come to a conclusion. Fungi are complex organisms and contain numerous compounds, some that may very well have medicinal properties. Unfortunately, what exactly is to be found in the numerous fungi supplements available is indeterminable since dietary supplements are not regulated as drugs and are not obligated to list their exact composition. What does, however, pop out of all the studies is that incorporating mushrooms into the diet is a good idea. But unless you are an expert, picking your mushrooms in the grocery store is the way to go rather than hunting for them in the wild. Mushrooms may or may not be able to cure, but they can certainly kill. As little as half a death cap (*Amanita phalloides*) can destroy the liver. "The Case of the Murderous Mushroom" would be an enticing title for a Sherlock Holmes story. Somebody should write it.

WHAT IF DR. KELLOGG HAD WATCHED *SEINFELD*?

Dr. John Harvey Kellogg would have enjoyed "The Contest" and "The Abstinence," two classic *Seinfeld* episodes. In "The Contest," Jerry, George, Kramer, and Elaine wager on who can remain master of their domain for the longest time, and in "The Abstinence," George and Elaine experience the consequences of temporary celibacy. Why would Dr. Kellogg who ran the Battle Creek Sanitarium 100 years before *Seinfeld* aired on television have enjoyed these plot lines? Because he staunchly believed that sexual activity, self-satisfaction in particular, had a negative impact on health.

Cancer, urinary disease, epilepsy, sleeplessness, loss of vitality, vision impairment, and oddly atrophy of the testicles and impotence were all due to engaging in pleasures of the flesh. The antidote? A bland vegetarian diet that Kellogg claimed reduced sexual stimulation. Interestingly, he was also one of the first to note that it took huge amounts of grain to produce small amounts of meat. The same grain, he maintained, could just as well have been used as food, a point environmentalists are making today. Dr. Kellogg also recommended using soy as a source of protein rather than meat, an idea that is well accepted today. The good doctor could not be accused of not following his own advice. He maintained a vegetarian diet and never consummated his marriage with Mrs. Kellogg, but the couple did foster forty-two children, eventually adopting eight of them.

Undoubtedly, among the general public, Dr. Kellogg is best known for having developed the flaked breakfast cereal. However, the notion that he invented cereal flakes to curb the sexual appetite is a myth. The various cereals developed in the kitchen of "The San" were supposed to be easily digested, and their low protein and high fiber content were said to change the intestinal flora for the better. Dr. Kellogg was vehemently opposed to adding sugar to the cereal he and his brother Will had formulated out of wheat, oats, and corn, and that eventually caused a falling out with Will who was more interested in profits than health. Will went on to found the Kellogg cereal company that made a fortune by selling sugary cereals, much to the horror of his brother. And that was before the company produced Froot Loops, Frosted Flakes, or Corn Pops.

With his view of the importance of the intestinal flora, Dr. Kellogg could be called a visionary. Today, researchers crank out almost daily publications on the role of the microbiome in health. Not only did Dr. Kellogg believe that the fiber in his cereals — today we would refer to it as a *prebiotic* — was conducive to good health, he also advocated the use of yogurt to balance the intestinal flora. Without using the term, he was pushing probiotics. He even had the idea that the live bacteria in yogurt could be best introduced into the colon through enemas. Although this has not been put to a test, fecal matter containing live bacteria and transplanted through the rear portals have been shown to be effective in treating some intestinal diseases.

Given his penchant for yogurt, Dr. Kellogg would likely have taken an interest in another *Seinfeld* episode, "The Non-Fat Yogurt." Jerry, George, and Elaine can't believe how good a frozen yogurt that Kramer has invested in tastes and assume that since it is advertised as "fat-free," they can eat as much as they like. Only when they start gaining weight do they realize that the advertising is false. I think Dr. Kellogg would have liked the idea of eating yogurt, and with his lack of interest in profits, would have been up in arms about false advertising.

Dr. Kellogg would also have been in line with another scene in "The Abstinence" in which Kramer, upon seeing smokers being forced to smoke outside a restaurant, is inspired to convert his apartment into a smoking lounge. But he then experiences the effects of second-hand smoke, which makes his face leathery, and enlists a lawyer to sue the tobacco company for ruining his good looks. Dr. Kellogg was an early opponent of tobacco smoking, long before Richard Doll conclusively linked smoking to lung cancer in the 1950s. He believed, correctly, that tobacco was not only physiologically damaging but also presented an economic burden for society. Kellogg also claimed, again correctly as we now know, that alcohol has a deleterious effect on the brain, the digestive system, and the liver.

I think Dr. Kellogg would have been a bit ambivalent about the episode in which George consults a "holistic healer," hoping to avoid surgery to have his tonsils removed. After telling George that he should have been born in August instead of April, the healer tells him that he must never take a hot shower. Dr. Kellogg was a big believer in hydrotherapy, especially reducing the pulse with a cold-water bath. But drinking a tea containing "cramp bark," "cleavers," and "couch grass" while sitting under a pyramid would not have sat well with Dr. Kellogg, who was a trained and competent surgeon and was not an advocate of herbal remedies. He would certainly not have agreed with the healer's allegation that the medical establishment just wants to keep people sick so they can reap more profits.

One episode that Dr. Kellogg would have found troubling is "The Mango," in which George's virility is boosted by eating a mango, a fruit that has an unwarranted reputation as an aphrodisiac. Given his views about the evils of sexual activity, the father of the corn flake would not have been amused!

While in many ways Dr. Kellogg was ahead of his time, he had a number of strange ideas in addition to his belief that carnal acts should be reserved for procreation. He was a disciple of Horace Fletcher, a dubious health expert who advised people to chew each bite of food at least forty times before swallowing. Kellogg also invented a vibrating chair that shook up to sixty times a second with the apparent goal of stimulating the bowels. Then there was a device that administered a mild electric current to the skin, which Kellogg claimed could treat lead poisoning, tuberculosis, obesity, and, when applied directly to the patient's eyeballs, a variety of vision disorders. And we won't even mention the cage he invented to be placed over young boys' genitals to prevent the great evil.

PSYCHEDELICS AND BICYCLE DAY

Unless you are a researcher or user of psychedelics, April 19 will not have any special meaning. However, should you fall into one of those categories, you will likely recognize the date as Bicycle Day, launched in 1985 by psychology professor Thomas Roberts of Northern Illinois University to commemorate the first ever intentional "acid trip" taken by Swiss chemist Albert Hofmann. And yes, a bicycle really was involved in that historic trip.

In 1938, Hofmann, working for the Sandoz pharmaceutical company, was researching compounds found in the ergot fungus that was known to contaminate grains. Consuming such tainted grains causes a condition known as St. Anthony's Fire, characterized by convulsions, vomiting, and hallucinations. But extracts of the fungus had also found a role in controlling excessive bleeding after childbirth. Hofmann was interested in isolating the active ingredients from the fungus and possibly synthesizing safer and better analogues. Indeed, he managed to convert one of the components of ergot into Methergine, a drug that was then used throughout the world to control blood loss after childbirth. This whetted his appetite for synthesizing other potentially useful ergot derivatives, one of which turned out to be lysergic acid diethylamide, or LSD. No useful effects were found, but some of the animals given the drug became excited. That did not excite Sandoz, and any further research on LSD was shelved.

For five years Dr. Hofmann harbored a lingering suspicion that his "problem child," as he would refer to LSD in his 1979 book, might have a use after all. In 1943, he decided to have another go at synthesizing the compound. That afternoon, as he later recounted, he was forced to stop work due to being seized by a "peculiar restlessness and mild dizziness." On arriving home, he experienced a "stream of fantastic images of extraordinary plasticity and vividness accompanied by an intense, kaleidoscope-like play of colors." Could this have been due to an accidental ingestion of LSD? There was only one way to find out. Dr. Hofmann would become his own guinea pig.

Surely only a very tiny amount of LSD could have been ingested by chance, so Hofmann decided to try a quarter of a milligram, the smallest dose that based on his research with ergot derivatives might be expected to have an effect. And did it ever! In less than an hour he began to experience unrest, visual disturbances, and dizziness. Time to leave the lab! His usual mode of transport was a bicycle, and he now hopped on and headed home. This would become the epic journey eventually commemorated by celebrating Bicycle Day. It was on this trip that Hofmann experienced his "field of vision swaying and seeing objects distorted like images in curved mirrors." A physician was called as soon as he got home, to whom he reported "a heavy feeling in the head, vertigo, and sounds being transformed into optical effects with every sound evoking a corresponding colored hallucination."

Dr. Hofmann described his adventure to colleagues, some of whom then volunteered to see if the effects were reproducible. They were, at doses even much smaller than the original quarter of a milligram. This set Hofmann wondering about mental illness, particularly schizophrenia, since some of its symptoms seemed to resemble the effects of LSD. Could it be that traces of some psychoactive substance produced by the body itself might produce psychic disturbances? Not being a physician, he did not explore this further, but he noted that part of the LSD molecule resembled the neurotransmitter serotonin.

Dr. Humphrey Osmond, a British psychiatrist who settled in Canada made a similar observation. In 1952, he noted a similarity in molecular structure between the hormone adrenaline and mescaline, the hallucinogenic component of the peyote cactus that had a long history of ceremonial use by Indigenous people of Central America. He also found a similarity between the effects of mescaline and symptoms of schizophrenia and seconded Hofmann's hypothesis that this mental disease might actually be a form of self-intoxication. As it later turned out, Osmond was on the right track. The culprit wasn't adrenaline but a chemically similar compound, dopamine. Overproduction of dopamine, one of the body's natural neurotransmitters, can cause some of the symptoms of schizophrenia, and indeed, the first effective medication to treat the disease, chlorpromazine, introduced in 1953, worked by blocking the action of dopamine.

Osmond's linking the effects of mescaline to some of the symptoms of mental disease led him, like Hofmann, to engage in some self-experimentation. He wrote about his mind-expanding and mystical experiences with mescaline as well as with seeds of a species of morning glory that were later found by Hofmann to contain lysergic acid derivatives. For such compounds, Osmond proposed the term *psychedelic*, from the Greek words for *mind* and *manifest*, with the idea being that they can develop the unused potentials of the human mind.

Aldous Huxley, the writer who had become famous for his book *Brave New World*, heard about Dr. Osmond's exploits and wrote to him lamenting how contemporary education was constricting students' minds. He wondered if mescaline could be used to open them up. Could Dr. Osmond give him some mescaline to try? Under Osmond's supervision, Huxley took mescaline and ended up writing a book, *The Doors of Perception*, that documented his "trip" and brought psychedelics to the attention of the public. Huxley had actually suggested his own made-up word, *phanerothyme* from the Greek for *spirit* to describe such chemicals and sent Osmond a rhyme: "To make this mundane world sublime, just half a gram of phanerothyme." Osmond retorted with, "To fathom Hell or soar Angelic, you'll need a pinch of psychedelic." Quite insightful. Let me add one of my own, with which I think Dr. Hofmann, who died in 2008 at the age of 102, would agree. "If you take a mystical chemical, don't go riding a bicycle."

TIMOTHY LEARY'S JOURNEY FROM RESPECTED SCIENTIST TO PSYCHEDELIC CELEBRITY

The place? Room 1742, the Queen Elizabeth Hotel in Montreal. The date? June 1, 1969. The event? A recording of "Give Peace a Chance" by John Lennon and Yoko Ono during their celebrated "bed-in," a non-violent protest against war. Also heard on the recording are a number of guests John and Yoko had invited. Tommy Smothers plays guitar, poet Allen Ginsberg, comedian Dick Gregory, singer Petula Clark, and Timothy Leary vocalize along. Yes, that Timothy Leary. The former Harvard professor who became an icon of the counterculture movement with his promotion of psychedelics and his mantra of "turn on, tune in, drop out."

The *turn on* referred to LSD and psilocybin, mind-altering drugs that Leary claimed would "enable each person to realize that he is not a game-playing robot put on this planet to be given a Social Security number and to be spun on the assembly line of school, college, career, insurance, funeral, goodbye." Timothy Leary's path from respected scientist to psychedelic celebrity is an intriguing one.

Having just obtained a PhD from Berkley in psychology, Leary was hired by Harvard University in 1959. His early academic career, studying social relationships and psychotherapy, was standard and unremarkable. But that was all to change in 1960 with a vacation to Cuernavaca, Mexico. Sitting around the pool one day, a conversation with colleague Anthony Russo turned to the latter's experience on a previous trip during which he had eaten magic mushrooms. He had read about these in a 1957 article in *Life Magazine* entitled "Seeking the Magic Mushroom" by Robert Gordon Wasson, an amateur mycologist.

Wasson recounted how in 1955 he had convinced Maria Sabina, a traditional healer, to allow him to take part in an ancient Indigenous ritual using "sacred mushrooms." His description of the effects, particularly visions, sparked immense interest, including in Russo who then traveled to Mexico to try the fungi. He now wanted to repeat the experience, and with help from University of Mexico anthropologist Gerhardt Braun who had studied the use of hallucinogens in Mesoamerica and managed to procure some mushrooms. Russo, Braun, and Leary then embarked on a psychedelic trip that Leary would describe as "a classic visionary voyage from which he came back a changed man." He had learned more about psychology during the five-hour trip, he said, than he had in fifteen years of studying the subject as an academic. From that moment on, Leary's research would take on a new phase. The psychedelic effect, he concluded, needed further exploration.

Leary quickly learned that Wasson had sent a sample of the mushrooms he had consumed to Albert Hofmann, the Swiss chemist who had achieved fame by synthesizing LSD while working for the Sandoz pharmaceutical company. Did some component of the mushroom perhaps resemble LSD? Wasson wondered. Hofmann was able to identify psilocybin as the compound in the "sacred mushrooms" that was indirectly responsible for the psychedelic effect. In the body, psilocybin is quickly metabolized to psilocin, a compound that, like LSD, has a chemical similarity to the neurotransmitter serotonin. The hallucinogenic effect is due to the overstimulation of serotonin receptors on cells of the nervous system. Hofmann also was able to synthesize psilocybin which Sandoz then made available for medical research.

Back at Harvard, Leary shifted his research to the effects of psilocybin which at the time was totally legal. At first, he followed proper scientific methodology and investigated the possibility of aiding alcoholics with psilocybin and using it to rehabilitate released prisoners. The Concord Prison Experiment, in which Leary claimed that his use of psychotherapy combined with psilocybin reduced the rate of released prisoners returning to criminal activity from 60 to 20 percent, received extensive publicity. Follow-up studies contested Leary's results, but by that time he and colleague Richard Alpert had gone on to experimenting with the effects of psilocybin on human consciousness by administering it to volunteers. Issues began to be raised by other faculty members about the way this research was being conducted. Apparently, some disturbing side effects were not disclosed, and some of the research was carried out when Leary and Alpert were under the influence of psilocybin. There were also allegations that the duo were actively promoting the use of hallucinogens for recreational purposes and had been giving drugs to undergraduates. This did not sit well with the Harvard administration, and in 1963 both Leary and Alpert were dismissed.

There would be no more academic appointments in Leary's future, but his promotion of the use of psychedelics under controlled conditions "for serious purposes, such as spiritual growth, pursuit of knowledge, and personal development" would make him an icon for the counterculture movement of the sixties. In a famous interview with *Playboy* magazine in 1966, he curiously claimed that "in a carefully prepared, loving LSD session, a woman will inevitably have several hundred orgasms." He then went on to say that "your life before was a still photograph that with LSD comes alive, balloons out to several dimensions, and becomes irradiated with color and energy."

President Nixon wanted no part of any such ballooning and called Leary "the most dangerous man in America." The legal system did indeed treat him as a dangerous man, ridiculously sentencing him to thirty years in prison for possession of a small amount of marijuana. That conviction was overturned by the Supreme Court, but not before he had spent time in prison, including one stint in Folsom State Prison in a cell next to murderer Charles Manson. In the 1970s and 1980s, Leary spoke on campuses, wrote articles, and continued to describe his own experimentation with an array of psychedelic drugs. Being leery of the law, Leary was careful about promoting their use.

There is no question that the name of Timothy Leary has to appear prominently in any discussion of psychedelics. While some of his unorthodox experiments deserve rightful criticism, his claim of improved creativity and cognition with the use of small doses of psychedelics is now getting traction with researchers studying microdosing.

Today, Montrealers and tourists walking along a path on the city's Mount Royal can recall the epic day when Timothy Leary crooned along with John Lennon and Yoko Ono. The path features a sculpture of limestone slabs, each one engraved with the phrase "Give Peace a Chance" in one of forty languages. We need to bring some world leaders here to walk that path.

MONITORING FREQUENCY NONSENSE

Down the rabbit hole I went. But unlike Alice, who ended up in Wonderland, I ended up in Heartland. Specifically, Hilton's Heartland. And there, like Alice in Wonderland, I experienced strange adventures galore.

In this era of the internet, "down the rabbit hole" has taken on a new meaning. Start out on a research path and discover that it leads to some unforeseen destination from where other paths fan out and lead to other attention-grabbing sites. My journey started with a rather challenging question I was asked. "Is it true that you can raise the body's frequency with rose essential oil?" I had no idea what to make of this, since *body*, *frequency*, and *essential oil* did not seem to belong in the same sentence. So, what does one do these days? Plug the terms into a search engine to see what pops up. And up popped the name of Gary Young, founder of Young Living, a multi-level marketing company specializing in essential oils. Young, with no apparent relevant scientific background, had written a book, *Human Electrical Frequencies and Fields*, in which he described how his "patients" reported pain relief in their neck when an essential oil was applied to their feet. "I started to realize that there had to be other aspects and elements in the oils that had to be researched," he wrote.

And research he did. Young discovered the "Calibrated Frequency Monitor," a device "that measures the frequencies of essential oils and their effect on human frequencies when applied to the body." This was invented by Bruce Tainio, who, according to his obituary, was "an inventor and a student of energy and quantum physics that led him to invent several instruments which minimize environmental stress from electromagnetic frequencies." Using the Frequency Monitor, a description of which cannot be found, Young determined that a healthy body typically has a frequency ranging from 62 to 78 Hz, while disease begins at 58 Hz. Processed food, he found, has zero frequency while essential oils start at 52 Hz and go as high as 320 Hz for rose oil. He goes on to say, "I believe that the chemistry and frequencies of essential oils have the ability to help man maintain the optimal frequency to the extent that disease cannot exist." I have a slightly different belief. I believe talking about the frequency of the body and the frequency of essential oils doesn't even rise to the level of pseudoscience. It is pure gibberish.

Frequency is the number of times an event occurs per unit of time with a Hertz (Hz) being one event per second. For example, when the middle key on a piano is played, it causes a string to vibrate 262 times a second, and that vibration is described as having a frequency of 262 Hz. But talking about the human body having a frequency makes no sense. If we were vibrating at 60 Hz, we would certainly feel it. Assigning a vibrational frequency to rose oil is also meaningless. While bonds within a molecule do stretch and contract, molecules do not have a vibrational frequency. Furthermore, rose oil is not a single entity, it is composed of hundreds of compounds. The nonexistent frequency of the body cannot be increased by the nonexistent frequency of rose oil.

The Calibrated Frequency Monitor was not the only device in Young's bag of tricks. To support the frequency theory, he referenced Royal Rife's "frequency generator." Rife was a self-educated inventor who in the 1930s claimed to have found a way to neutralize various disease organisms with radio waves. These organisms, he said, have a "Mortal Oscillatory Rate" and can be destroyed by vibrating them with radio waves of a specific frequency. The American Medical Association maintained, and continues to maintain, that there is no evidence that Rife's rays can cure disease. Rife therapy devotees retort that there is a conspiracy in the medical community to keep an effective cancer cure from the public, and they continue to sell various Rife machines to patients. Rife himself never claimed to cure cancer.

I thought Rife merited a further look and that is how I ended up on the website of Hilton's Heartland, a "Natural Health & Wellness Center" in Texas. I learned that "everything in the physical world is comprised of energy, and energy vibrates at different levels. The vibratory rate of a tomato, for instance, is different from the vibratory rate (or frequency) of a cucumber, just as the frequency of a bacteria is different from the frequency of a fungus." A mishmash of meaningless words. But there is more. "Frequencies specific to all different forms of disease — including influenza, cancer, Parkinson's disease, multiple sclerosis, and an immense number of other conditions have been identified and captured in Rife Technology." Of course, this technology is offered to clients, but there is a disclaimer. "The Center's employees do not practice medicine and do not examine, diagnose or treat, or offer to treat or cure or attempt to cure, any mental or physical disease, disorder or illness, or any physical deformity or injury." So, if there is no diagnosis or treatment for anything being offered, why would anyone visit Hilton's Heartland? Because in spite of the disclaimer, various treatment are indeed offered!

People suffering minor arthritic pain can stand on a BioMat and enjoy "remarkable benefits from a combination of far infrared rays, negative ions and the conductive properties of amethyst channels." Or, starting at $175, they can undergo computerized Recalibration Sequencing using "interrogatory biofeedback." There is also "detoxification" with homeopathic or herbal remedies, as well as acupressure, lymphatic enhancement therapy, light touch therapy, and cold laser therapy "to rebalance body systems and rejuvenate at a cellular level." Cranio Sacral Therapy "permits improvements in brain and spinal cord function which, [*sic*] increases the health of your entire body, enhances your resistance to disease and assists in a return to optimal function." And yes, they offer essential oils but mercifully make no mention of any frequencies.

The founder and owner of the wellness center describes herself as a Doctor of Naturopathy, Doctor of Pastoral Science, Master Herbalist, Certified Natural Health Practitioner, and Cranio Sacral Therapist.

She seems to have yet another talent. She excels at making the jumbled word salads that I found at the bottom of the rabbit hole.

WILHELM REICH AND THE ORGONE ACCUMULATOR

Gwyneth Paltrow seems to be fond of tantalizing us with tales of her bodily orifices. It may be the saga of drinking goat milk for eight days to rid her body of non-existing parasites, or exercising her genitals with jade eggs, or, the latest, insufflating her rectum with ozone gas. Exactly why she pumps a potent oxidizing agent up her rear isn't clear. "It has been very helpful," the Goddess of Goop says, without any mention of what such rectal inhalation may be helpful for. "Nothing," would be the scientific consensus. Yes, there are various alternative practitioners who claim that ozone can treat conditions ranging from back pain and arthritis to, most disturbingly, cancer. Some administer the gas intravenously, others apply ozonated oils to the skin. There are even ozonated waters to drink and saunas in which the patient sits in a cabinet filled with the gas. Thankfully with the head outside the cabinet since inhaling ozone can damage the lungs. All these treatments have one thing in common. They lack evidence of efficacy.

Ozone saunas may be ineffective, but they do expose the subject to ozone, a very real gas. The same cannot be said for people who sit in an Orgone Accumulator with hopes of boosting their sexual energy or treating some disease. The Accumulator is a wooden cabinet about the size of a telephone booth with a single chair inside. It is supposed to charge the body with energy by accumulating "orgone" from the atmosphere. A metal lining prevents the captured orgone from escaping. But there is a slight problem here. The evidence for the existence of the mysterious orgone is, well, zero. Nevertheless, its discovery, or rather invention, does make for a fascinating, albeit bizarre story.

Wilhelm Reich was born in 1897 in what was then the Austro-Hungarian Empire and went on to get a medical degree from the University of Vienna. He specialized in psychiatry and in the 1920s became part of Sigmund Freud's inner circle. The "father of psychoanalysis" was drawing attention at the time with his theory that neuroses are caused by a conflict between society's suppression of natural sexual instincts and the body's urge to express these instincts. The notion appealed to Reich who pushed the idea further to claim that neurotic symptoms could be alleviated by sexual gratification. Orgasm, he maintained, discharges excess biological energy that if allowed to build up fuels neurotic disorders. According to Reich, sexual misery in society was rampant and sufferers needed to be educated about the essential role of sexuality in life.

Reich's ideas about sex being healthy and restraint being unhealthy did not sit well with the medical establishment or with the fascists who were gaining power in the 1930s. The latter certainly did not like the insinuation that their odious beliefs emerged from sexual frustration. Reich recognized that his ideas had no future in Europe and left for America where he believed that the notion of a better orgasm curing society's ills would be more acceptable. Indeed it was. Especially after Reich introduced some pseudoscientific conjecture about the source of the energy that fueled the orgasm. It was an invisible substance he called *orgone*, which he claimed permeated the atmosphere from where it could be absorbed to vitalize the body and maintain health. But if allowed to build up excessively, it caused problems. Luckily, the excess could be released by sexual satisfaction.

How Reich came to formulate his ideas about the existence of orgone is somewhat of a mystery. At one point he described examining cells through a microscope and observing some blue particles that moved about energetically. Somehow he concluded these came from the air, but he never explained how these particles, which nobody else has seen, were the source of orgone energy, which nobody has ever measured. But the lack of evidence for the existence of orgone did not stop Reich from producing Orgone Accumulators. Sitting in one would cure disease and enhance the power of a future orgasm by endowing its users with "orgastic potency." There was more. Besides leading to genital utopia and miraculous cures, orgone could also produce rain. Reich built a bizarre machine that supposedly generated the elusive orgone and discharged it at the sky in a process he called "cloudbusting." It was a bust.

Reich's sexual hedonism and his claims about the wonders of the Orgone Accumulator gained traction with the media and brought his exploits to the attention of the Food and Drug Administration. In 1954, the FDA asserted that the Orgone Accumulator was a sham and claims for its effectiveness were false and misleading. An injunction was issued against further sales and any dissemination of literature that promoted the healing properties of orgone. When Accumulators continued to be sold in spite of the injunction, Reich was arrested and sentenced to prison where he died in 1957. But the Accumulator did not die with him.

Writers such as Norman Mailer, Jack Kerouac, and Allen Ginsberg, all pioneers of the Beat Generation, embraced the device. Then there was Sean Connery who, during his James Bond days, gave it a shot. As the story goes, he had some virility problems due to an inner conflict between himself and his James Bond persona. Whether he received any satisfaction from the machine is not known. Actually calling the contraption a machine is misleading because it had no moving or electrical parts of any kind. The only thing it delivered was a dose of placebo.

Woody Allen parodied the Orgone Accumulator in his 1973 film *Sleeper*, in which the hero, owner of a health food store, undergoes cryopreservation and wakes up in the twenty-second century. In this dystopic America, sexual ecstasy can be achieved in a cabinet called the *orgasmatron* without requiring any messy physical contact. A clear dig at Reich's Accumulator.

A few original Orgone Accumulators can be found in museums, but so far the cabinets have apparently escaped the attention of Gwyneth Paltrow. A modern version, perhaps with some blinking lights, would undoubtedly entice Goop's customers. Maybe the chair could even be fitted with an ozone infuser so that the sitter could enjoy the fictitious effects of the treatment while bathing in the mythical orgone.

CHOLENT AND THE MAILLARD REACTION

When it comes to taste, my mother's Hungarian version of cholent loses out to my mother-in-law's Romanian variety. Back when they were both still with us, I had plenty of chances to compare. Of course, at the time I didn't know anything about advanced glycation end products, or indeed how such AGEs may accelerate aging. If instead of taste, the competition were about health, my mother's version, which is more of a stew, would have defeated my mother-in-law's oven-baked cholent.

Cholent is a traditional Jewish dish that can be traced back to at least the ninth century. It was born out of dedication to follow the Old Testament's edict of observing Shabbat as a day of rest that precludes any work, including cooking. The idea was to assemble all the ingredients in a pot that was then brought to a local baker before sundown on Friday and placed in an oven where it cooked overnight. The cholent was ready to be picked up the next day so that the family could enjoy a hot Shabbat meal.

Over the years, cholent became a popular dish both among observant and non-observant Jews, with many variations of ingredients and cooking methods being adopted. Both of the versions to which I was treated had several different kinds of beans, barley, onions, garlic, carrots, celery, beef short ribs, vegetable oil instead of the traditional schmaltz (chicken fat), and, needless to say, paprika. My mother slow cooked the cholent on the stovetop, which means it simmered at 100°C, while my mother-in-law baked it in the oven at around 200°C. The high temperature delivered more flavor, and undoubtedly more AGEs.

Now about those AGEs. We have to take a trip back to 1912 when Louis Camille Maillard first described a reaction that takes place between the simple sugar glucose and amino acids. The adduct that forms under the influence of heat can then go on to engage in a host of other reactions leading to a large variety of flavorful, aromatic, and colorful compounds that are largely responsible for the taste, smell, and appearance of baked bread, roasted coffee, cocoa, grilled steak, beer, and, of course, cholent. Collectively, when it comes to cooking, these reactions are referred to as the Maillard reaction, paying homage to its first investigator. Louis Camille Maillard had a medical degree but preferred to practice chemistry. "Let us wish that medicine will never treat physico-chemical sciences as subordinate; they constitute its very foundation," he declared in a speech on the centenary of Louis Pasteur's birth. Pasteur, credited with connecting germs with disease, was a chemist.

In medical circles, the Maillard reaction is known as *glycation*, from the Greek word for sweet since it is based on the reaction of a sugar with an amino acid. The compounds that then form as a result of the initial reaction are termed *advanced glycation end products*, or *AGEs*, which brings us to the troublesome part of the Maillard reaction, its impact on health.

The first alarm was sounded in 2002 when Swedish food chemists Margareta Tornqvist and Eden Tareke discovered the presence of acrylamide, an AGE, in french fries, potato chips, and biscuits. That finding sent shock waves through the food industry because eight years earlier the International Agency for Research on Cancer (IARC) had declared acrylamide a "possible human carcinogen." Acrylamide forms as a result of a reaction between glucose and the amino acid asparagine, commonly found in many foods. The higher the temperature, the more likely the reaction. That is why in California, abiding by the state's controversial Proposition 65, french fry containers warn customers about the presence of acrylamide.

Apart from being formed in foods, AGEs can also form in the body. After all, we are full of sugars and amino acids, both individual ones and ones incorporated into proteins. The Maillard reaction is known to take place even at lower temperatures, albeit more slowly. A classic example is the reaction between glucose and lysine, an amino acid that is part of hemoglobin, the molecule in red blood cells that transports oxygen. This glycated hemoglobin is known as HbA1c, and its concentration in the blood reflects the average level of glucose to which red blood cells have been exposed during their life cycle of about three months, and therefore A1c is a measure of blood sugar control. This glycated product is not toxic, but other AGEs, such as acrylamide, have been associated with various conditions ranging from kidney disease and atherosclerosis to Alzheimer's disease and cancer.

When AGE-rich diets are fed to mice, they develop kidney disease and atherosclerosis. In humans, dietary AGEs correlate with circulating AGEs, as well as with markers of oxidative stress, a condition in which there are not enough antioxidants in the body to counter tissue-damaging free radicals. Further warnings about AGEs, acrylamide in particular, come from an analysis of data collected in the National Health and Nutrition Examination Survey (NHANES) in the U.S. between 2003 and 2014. Researchers (curiously in China with no connection to the U.S.) studied the results of the American survey and found a strong association between biomarkers of acrylamide, markers for inflammation, and cancer mortality. Besides cancer, inflammation is known to be a factor in various diseases.

While the accusations against AGEs being dietary criminals are not ironclad, enough red flags have been raised to warrant attempts to decrease our intake. How? A baked potato has one-tenth the AGEs of french fries, and even in the fries, AGEs can be reduced by 90 percent if potatoes are first soaked in water. A broiled steak has three times as many AGEs as braised beef, grilled chicken five times as much as poached, and a fried egg has fifteen times as much as a scrambled egg. A diet of vegetables, legumes, fruits, whole grains, with low-fat dairy and less meat will result in fewer circulating AGEs.

So how do I make my cholent? I go for taste and use my mother-in-law's recipe. I make it so rarely that I don't worry about the AGEs. But I am careful not to burn my toast, I look askance at potato chips and cookies, and I steam my vegetables. When it comes to cooking, "low, slow, and moist" win the day. Instead of grilled chicken, I go for slow cooked paprika chicken. That means less worry about aging due to AGEs.

NOT SO SWEET TALK ABOUT SUGAR

"That tastes really sweet," the native must have thought to himself as he chewed on the stalk of a plant he had just pulled out of the ground. The place? New Guinea. When? Some 8,000 years ago. And thus began the human love affair with sugar. It didn't take long to discover that sugar cane could easily be propagated by planting cuttings and domestication of the plant was under way. When Indian traders began to explore the Pacific islands, they learned about sugar cane and introduced it to India from where it spread to China. Somewhere along the way it was discovered that squeezing the sugar cane expressed a juice that if allowed to stand formed sweet crystals. By the first century, sugar cane mills were introduced in India and knowledge of sugar spread westward.

The Greeks and Romans learned about sugar by the fourth century and used it mostly as medicine. In the Middle East, Arab farmers learned to cultivate sugar cane and became proficient at extracting its sugar content. They were also the first to combine sugar with other commodities such as almonds to make marzipan. The first Europeans to learn of sugar were the Crusaders who brought the sweet substance back from Jerusalem in the eleventh century.

Sugar cane cannot be grown in the cooler European countries, but when the Spanish colonized the Canary Islands they found it to be an ideal place to grow the plant. Sugar mills were set up and Indigenous people were enslaved to run them. The Portuguese did the same on the island of Madeira. On his second voyage to America, Columbus took sugar cane cuttings from the Canaries to the island that is now Haiti and the Dominican Republic and thus planted the seed for the sugar industry in the Caribbean. The Portuguese followed suit and established sugar plantations in Brazil. They came not only with seedlings but with muskets that outmatched the bows and arrows of the natives, who were enslaved. Sugar production increased and Europe now had sugar from America as well as from the Canaries and Madeira. Curiously, it was regarded not only as a sweetener but as a medicine as well.

In the sixteenth century, German physician and botanist Jacobus Theodorus Tabernaemontanus wrote, "Nice white sugar from Madeira or the Canaries, when taken moderately cleans the blood, strengthens body and mind, especially chest, lungs, and throat, but it is bad for hot and bilious people, for it easily turns into bile, also makes the teeth blunt and makes them decay. As a powder it is good for the eyes, as a smoke it is good for the common cold, as flour sprinkled on wounds it heals them. With milk and alum it serves to clear wine. Sugar water alone, also with cinnamon, pomegranate and quince juice, is good for a cough and a fever. Sugar wine with cinnamon gives vigor to old people. Sugar candy has all these powers to higher degree." At least he got the bit about blunting the teeth and making them decay right.

With the introduction of coffee and chocolate to Europe, sugar as a sweetener became even more desirable. As demand increased, so did the need for manpower on plantations and the mills. This was filled by African slaves, some half a million of whom were shipped to the New World in the seventeenth century. By the eighteenth century, movements to abolish slavery were growing and American abolitionists promoted the use of maple syrup instead of sugar from the Caribbean. But slavery continued until 1866 in the U.S. and 1888 in Brazil.

Sugar cane is not the only source of sugar. The high sugar content of beets was apparently first noticed in 1747 by Andreas Marggraf, a German apothecary, who had been using white beets as a laxative and was struck by the sweet taste. One of his students, Franz Karl Achard, became interested and planted some beets to see which variety produced the most sugar. By 1802, a pilot sugar beet refinery had been built in Prussia, but the sugar it produced could not compete in price with that produced from sugar cane. Then along came the Napoleonic Wars and the British blockade of French ports. That is when Benjamin Delessert came to the fore and developed an efficient method of producing pure sugar by clarifying the juice squeezed from sugar cane. The secret was to pass the juice through charcoal, a material with an amazing property to filter impurities. For this, he was awarded the Cross of the Legion of Honour by Napoleon who reportedly took the Cross from around his own neck and gave it to Delessert. Soon forty beet refineries were supplying France with sugar, and the number expanded to 250 by the mid-1920s. Sugar went from being a luxury item to a widely available commodity. Today, about 20 percent of the world's sugar is produced from beets.

After cereals and rice, sugarcane is the world's most valuable crop. It has yielded commercial profits galore, but it has also plied the world with empty calories that have contributed significantly to the epidemic of obesity and its companion ailments, namely heart disease, cancer, and diabetes. Critics have made the point that no other crop occupies as much land and uses as many resources for so little benefit to humanity as sugar cane.

That isn't totally correct. In Brazil sugar is fermented into ethanol and cars are designed to run either on ethanol, gasoline, or a mix. The residue left after sugar cane has been crushed, known as *bagasse*, can be burned for energy to power the sugar mill, with excess being transferred to the electrical grid. As far as carbon dioxide production goes, sugar cane absorbs more when it grows than it releases when burned. Bagasse can also be converted into pulp to make paper and boxes as well as particle board for furniture. It can even be processed into soluble fiber that can be added to food to boost fiber intake.

Undoubtedly excess sugar consumption is a problem, but there is one sugar myth that can be put to rest. There is no "sugar high," and sugar does not cause children to be hyperactive. Still, cutting back is good advice. Do we really need to pump children with Cap'n Crunch, half the weight of which is sugar? Let them eat Fiber One.

SEEING RED WHEN IT COMES TO FOOD DYES

Do they affect children's behavior? Are they carcinogenic? Should they be banned? Those are questions that have been raised about a family of chemicals commonly referred to as *food dyes*. There are some contentious issues here to be sure. But one aspect of these color additives is not contentious. They have a colorful history!

As early as 1500 B.C., Homer in his Iliad described the use of saffron prepared from the stigma of the *Crocus sativus* flower to color food, a practice also adopted by the ancient Egyptians. Since the Egyptians were very familiar with numerous mineral pigments such as iron and lead oxides, and also knew how to extract red dyes from madder root or the crushed bodies of kermes insects, it is likely that some of these colorants found their way into food.

It wasn't until the Middle Ages that the first controversy about food coloring cropped up. Bread made from refined white flour came to be favored over coarse, dark bread because it looked to be more "pure." White bread was more expensive to produce, so some bakers took the short route and added chalk (calcium carbonate) or lime (calcium oxide) to dark bread to lighten it. This did not go down well with King Edward I (1272–1307) who introduced the first law dealing with food adulteration. "If any default shall be found in the bread of a baker," the king decreed, "let him be put upon the pillory and remain there at least one hour in the day."

Nevertheless, profits from the whitening of bread were seductive enough for the prospect of being pilloried for an hour not to be enough of a deterrent. The public was blissfully unaware of being deceived until 1820 when German chemist Friedrich Accum who had settled in England published his "Treatise on Adulteration of Food and Culinary Poisons." By this time, bakers had expanded their repertoire and were also adding alum (potassium aluminum sulphate) to bread, but they were not the only ones to engage in adulteration. Milk was commonly tinged yellow with lead chromate, and candy makers colored their sweets with copper, mercury, and lead compounds. Accum went as far as publishing the names of those selling the toxic products, but his exposé was drowned out by a public smear campaign against him organized by the adulterers.

A new twist in the thread of the history of food dyes, one that would lead to controversies that dominate to this day, was the discovery in 1856 of the first synthetic dye by William Henry Perkin. Prior to Perkin's accidental discovery of mauve in a futile quest to synthesize quinine, anyone wishing to color food was forced to rely on nature's palette. The key to Perkin's synthesis was the use of aniline derived from coal tar, a waste product of the production of coal gas and coke. Chemists were quick to capitalize, and soon produced a whole range of "coal-tar colors" that were embraced by the food industry since they were cheap to produce, were free of the acute toxicity of metal salts, and produced vibrant colors with very small doses. By the turn of the century, some eighty synthetic dyes were used to color food in a totally unregulated fashion.

Then came a pivotal moment. Dr. Harvey Wiley was appointed chief of the Bureau of Chemistry at the U.S. Department of Agriculture, the forerunner of the FDA. Wiley was disturbed by the lack of regulations with respect to substances added to food and became a champion of reform. He was the force behind the passage of the 1906 Pure Food and Drugs Act that for the first time imposed criminal penalties on selling "adulterated" food. Dyes were of particular concern because of their ability to conceal damage and make inferior foods more appealing. Wiley commissioned a study of the food dyes being used and concluded that only seven of these were safe. Over the years, some of these were eliminated and others were added as more and more research came to light. The difficulty of establishing safety is highlighted by the different lists of approved color additives in Europe, Canada, and the U.S.

Red Dye No. 2, also known as amaranth, was one of the synthetic dyes approved by Wiley and became a favorite of the food industry. Then in the 1950s, the first sign of trouble appeared when a study reported breast tumors in female rats fed the dye. In response, the FDA ordered follow-up studies in rats and mice. The results were not corroborated, but then came word that a couple of Russian studies had found intestinal tumors in rats. Even though experts judged these to be of poor quality and unreliable, FDA ordered another test on rats that seemed to indicate an increase in tumors in females. Critics claimed the experiment had been bungled, but still, enough confusion had been created for FDA to remove Red No. 2 from the market. Although this dye was never used by M&Ms, the company decided to remove the red candies out of concern that the public may choose to avoid all red dyes. This triggered a ridiculous conspiracy meme that the red M&Ms had aphrodisiac properties and that workers were stealing them for their nocturnal pleasures.

In 2007, synthetic color additives again came under scrutiny following publication of a study conducted by the University of Southampton and published in *The Lancet*. This six-week study was commissioned by the U.K. Food Standards Agency (FSA) and was intended to investigate whether certain color additive mixtures and the preservative sodium benzoate when consumed in a beverage cause hyperactivity in children. The Southampton Six, as the evaluated dyes came to be called, were Quinoline Yellow, Ponceau 4R, Azorubine (carmoisine), Allura Red (Red No. 40), Tartrazine (Yellow No. 5), and Sunset Yellow (Yellow No. 6). The first three have never been allowed in Canada or the U.S.

Based on subjective evaluations by parents and teachers along with administered tests, the researchers concluded that the dyes affected behavior. The study received extensive media attention and prompted the European Union to require warnings on foods that contained these dyes stating, "May have an adverse effect on activity and attention in children." To avoid labeling foods with this warning, manufacturers chose to replace the dyes with natural colorants. Neither Canada nor the U.S. require any sort of warning, but any synthetic dyes used must be identified by name on the label.

A current concern is about Red No. 3, or erythrosine, one that was originally approved by Harvey Wiley, and consequently by the 1906 legislation. It was also approved "provisionally" in cosmetics until the 1990s when studies showed that rats fed large doses developed thyroid tumors. This caused Red No. 3 to be then banned from cosmetics, but curiously it was allowed in foods because of a quirkiness in FDA procedures that makes it difficult to ban a substance that is on a permanently rather than a provisionally approved list.

The Center for Science in the Public Interest (CSPI) in the U.S. has filed a petition with FDA to initiate procedures to ban Red No. 3 because of the cancer connection. CSPI also points out, correctly, that accumulating evidence affirms that synthetic dyes can cause behavioral problems in some children and suggests replacement by natural dyes derived from the likes of paprika, cabbage, turmeric, and beets. Many companies are already doing this, which is commendable since synthetic dyes serve only a cosmetic purpose. So, why take any chance, even if the evidence of risk is inconclusive? Avoiding dyed foods is not a bad idea in any case, given that dyes are usually found in processed foods of questionable nutritional quality. Maybe seeing red when it comes to synthetic reds is justified.

THE BRAINY SCIENCE OF LECITHIN

I wish I could eat fish, but I can't. Fish allergy! That means no commercially made Caesar salad since today's versions contain anchovies. The original dressing did not have anchovies per se, but did have Worcestershire sauce which is flavored with anchovies. It was a blend of olive oil, lemon juice, garlic, salt, spices, Worcestershire sauce, and Parmesan cheese with the yolk of an egg. The latter is the key to producing a homogeneous product. Egg yolk contains lecithin, a term that refers to several closely related compounds capable of acting as emulsifiers, substances that can prevent oil and water from separating. This is due to their molecules having one end that is attracted to oily substances and the other to water.

So, if I want to enjoy a Caesar salad, I have to make my own dressing, which isn't that hard to do. Just whisk together all the ingredients except for the olive oil, which is then slowly added with continued mixing until the lecithin in the egg yolk does its job and produces a smooth consistency.

Lecithin can also be found listed as an ingredient in many commercial salad dressings, as well as in cooking sprays where its role is to prevent food from sticking to the pan. It is also a common ingredient in chocolate where it serves to reduce the viscosity of the molten product, making it easier to pour into molds. As a bonus, lecithin also prevents the dreaded "fat bloom" that occurs when cocoa fat separates and forms white streaks on the surface of the chocolate. For such uses, lecithin is obtained from soybeans as a byproduct of oil extraction. And you know where else lecithin is to be found? In our brains!

The ancient Egyptians didn't attach much importance to the brain. They didn't think it would be of much use in the afterlife and removed it by inserting special hooked instruments up through the nostrils to pull out bits of brain tissue before mummifying a body. The heart, they believed, was the center of a person's being and intelligence and could come in handy in the hereafter. It was therefore left in place. Aristotle also believed that the mind resided in the heart, but Hippocrates got neuroscience on track by arguing that the brain was the seat of thought, sensation, emotion, and cognition. Of course, he had no idea of how the brain functioned, of the intricate chemistry involved. Indeed, it would take another 2000 years until German physician Johann Thomas Hensing would take a stab at determining the chemical composition of the brain. Believing that the brain was "truly the throne of the soul and the abode of wisdom," he undertook a chemical analysis which in those days consisted of heating a sample to drive off volatiles, including water, and studying the solids left behind.

In 1669, German alchemist Hennig Brandt made a monumental discovery when in a search for the elixir of life he boiled a vat of urine to dryness and was stunned to see the residue glow in the dark. He managed to isolate a substance that he named *phosphorus* from the Greek word for *light-bearing*. By the early eighteenth century, word about this curious substance had spread, and Hensing recognized its presence in the ashes left after combusting brain tissue. An exciting discovery to be sure, since a light-emitting substance in the brain suggested it may be the spark that gives rise to thoughts.

Hensing did not go beyond demonstrating that the brain contains phosphorus, but French physician Antoine-Francois Fourcroy who had developed a fascination for chemistry did. Instead of subjecting fresh brains to heat, he introduced solvent extraction procedures, and using alcohol he isolated a substance from a human brain he described as a "greasy oil." Fourcroy mentored a number of young students, one of whom, Nicolas-Louis Vauquelin, put two and two together and showed that the greasy oil contained phosphorus, although he was unable to determine the exact composition of the oil. Then along came pharmacy professor Theodore-Nicolas Gobley who extracted egg yolk with boiling ether and isolated a viscous substance that was identical to the oily liquid Vauquelin had described in extracts of brain tissue. He named it lecithin from the Greek for egg yolk. But he did more. By 1874, Gobley had determined the molecular structure of lecithin and showed that it was a phospholipid, consisting of fatty acids, phosphoric acid, and a small molecule called choline bound together.

Just why egg yolks and brain tissue contain phospholipids was not clarified until the twentieth century when these compounds were found to be critical components of the membranes that separate and protect the interior of cells from the outside environment. Since cell membranes control what substances can enter and exit cells, their integrity is critical for the maintenance of health. Our bodies construct these phospholipids from fats, phosphorus, and choline, all sourced from the diet. Obviously, lecithin, found in foods such as eggs, meat, soy, and corn, can furnish these nutrients, but questions have also been raised about there being another role for the substance. Producers of dietary supplements point out that lecithin is a source of choline, which the body needs to produce the neurotransmitter acetylcholine, which is lacking in Alzheimer's disease. There are also claims that lecithin makes for healthier hair and skin, improves liver function, enhances memory, relieves arthritis, and lowers cholesterol. The only one of these with any evidence is the effect on cholesterol, but there is a but. Bacteria in the gut can convert choline from lecithin into trimethylamine oxide (TMAO) which has been implicated in the formation of plaque in coronary arteries.

Besides the risk-benefit ratio of lecithin supplements being uncertain, there is the usual problem of supplements not containing what the label says. Some lecithin supplements have been found to contain only fatty acids. In any case, I'm not concerned about the lecithin from the egg yolk in my Caesar salad. It even spawned this hopefully interesting trip through history and the kitchen. By the way, Caesar salad has nothing to do with Julius Caesar. It was the brainchild of chef Caesar Cardini, who introduced it in 1924 in his Tijuana restaurant. Although I suppose, if Julius ever ate a salad, it was "Caesar's salad."

THE POPE, LIONEL MESSI, AND YERBA MATE

Pope Francis used to do it. Lionel Messi does it. And so do millions of South Americans. They regularly dip their bombilla into a brew of yerba mate, traditionally contained in a calabash gourd. The bombilla is a type of drinking straw with one end closed except for a series of perforations designed to filter out bits of leaves and stems. Those bits of leaves and stems come from the *Ilex paraguariensis* tree, native to the area that eventually became Paraguay.

The Indigenous Guarani people living in the region were the first to add hot water to the heat-dried chopped leaves of the tree. The infusion, thanks to its content of caffeine, theobromine, and theophylline, had a stimulant effect that added a bit of a lift to an arduous life. Spanish colonists who arrived in the sixteenth century learned about yerba mate from the Guarani and quickly adopted the habit of frequently sipping the beverage. The local governor, Hernando Arias de Saavedra, claimed they were indulging too vigorously in what he believed was an unhealthy habit and sought to ban the beverage.

Jesuit missionaries who arrived in the seventeenth century with the goal of converting the natives to Catholicism at first also looked on the yerba habit as a vice, but then realized that the cultivation of the trees afforded a way to make the missions self-sufficient and profitable.

Harvesting the leaves from trees that grew in the wild was a difficult process, and the Spaniards had previously tried to cultivate them but without success. The Jesuits, however, found the secret, which supposedly was to pass the seeds of the tree through birds before planting. In any case, they did manage to cultivate the trees and built a profitable business selling yerba mate throughout South America. It never made it to Europe where cacao and coffee had already captured the market.

In the eighteenth century, rulers in Portugal, France, and Spain became wary of the influence that Jesuits had on the population. As part of their mission, Jesuits had established educational institutions that promoted a fierce loyalty to the Pope, whom the rulers saw as a foreigner intent on meddling in their internal affairs. This resulted in the expulsion of the Jesuits from a number of countries, including Spain and its colonies. Yerba plantations became neglected and the widening of the habit of drinking yerba mate slowed until the 1890s when farmers in Argentina, Uruguay, and Paraguay discovered the profitability of cultivating the trees. By the nineteenth century, people across South America were imbibing regularly, often socializing with a gourd of yerba mate being passed around, refilled with hot water after each individual had a turn sipping through the bombilla.

As one might expect with such a practice becoming so popular, scientists began to take an interest in its pros and cons. The first red flag appeared when epidemiologists noted an increased incidence of cancer, particularly of the esophagus and oral cavity, in countries with a high consumption of yerba mate. One theory is that this is due to tissue damage caused by the frequent consumption of a very hot liquid. The International Agency for Research on Cancer (IARC) deems this important enough for hot yerba mate to be categorized in its group 2A, reserved for substances that are "probably carcinogenic to humans." Another candidate for the cancer risk is the presence of carcinogenic polycyclic aromatic hydrocarbons (PAHs) that have been detected in the leaves as a result of exposure to smoke from burning firewood during the roasting of the leaves. These are the same compounds that are of concern in barbecued foods, but being mostly insoluble in water, only trace amounts end up in yerba leaf extracts.

Researchers have also explored the potential benefits of yerba mate. Caffeine is of course a known stimulant, but the amount in a serving of yerba is less than that in a cup of coffee. There are a number of polyphenols with antioxidant properties, chlorogenic acid being a prime example, that can be extracted from the leaves and appear in the bloodstream after drinking the beverage. Whether this has any clinical benefit is questionable. But if one believes in the benefits of polyphenols, there are much better sources available than yerba mate.

There have been some suggestive studies, mostly in animals, about weight loss, reduced formation of advanced glycation end products (AGEs) as seen in diabetics, as well as anti-inflammatory and even anticancer effects. None of these are compelling enough to prompt a regular regimen of drinking yerba mate. As far as the canned yerba beverages that have appeared in North America, mostly targeting "wellness" seekers with claims of "rejuvenating," "energizing," "brain boosting," and "detoxifying" properties, the hype drowns out the science.

Pope Francis, being an Argentinian, was fond of yerba mate and said he drank it every day for nerve problems. The pontiff reportedly also tried another South American beverage, a more controversial one. Coca tea! Before landing at the world's highest international airport in Bolivia, he drank a cup to ward off altitude sickness. Coca is the plant from which cocaine is extracted, but the amount of cocaine in a cup of tea brewed from the leaves is barely enough to produce mild stimulation. It is, however, enough to be detected in the urine which has led to athletes being disqualified from competitions. Although there is no scientific evidence for reducing the effects of altitude sickness, the tea is nevertheless consumed for this purpose in the Andes regions, particularly by tourists bent on scaling the mountains to reach Machu Pichu in Peru.

Francis was not the first Pope to have a connection with coca. In the late nineteenth century, Leo XIII actually endorsed Vin Mariani, a concoction made by steeping coca leaves in Bordeaux wine. He even awarded it a Vatican gold medal! Pope Francis would likely not have done that. Why? Because before entering the priesthood, he graduated from a technical school with a degree in chemistry! He would have known that alcohol is a good solvent for cocaine and that therefore coca wine would contain a significant amount of the potentially harmful chemical.

While there may be issues with beverages infused with coca, there are none with trying yerba mate. During Francis's pontificate, you could buy a stainless steel bombilla decorated with his image. I should have bought one. They are now collector's items.

KING GEORGE'S PURPLE URINE

You wouldn't think that King George III had anything in common with vampires. It is not that the king who reigned from 1760 to 1820 believed in the undead rising from their graves to suck the blood of the living, at least as far as we know. What links King George to vampires is a disease known as porphyria, from which he is said to have suffered, and which also has been proposed as having given rise to the myth of the vampire. While cogitations about King George, vampires, and porphyria have circulated widely in the media, including in the medical literature, the scientific underpinnings of the associations are highly suspect.

Porphyria encompasses a group of eight genetic disorders characterized by the buildup of molecules called porphyrins in the body. These are the building blocks of heme, which in turn is incorporated into hemoglobin, the molecule in red blood cells that binds oxygen and transports it to cells. In the porphyrias, there is a malfunction in one of the steps in the body's synthesis of heme, so porphyrins accumulate and cause the symptoms of the disease. These can vary depending on which specific enzyme needed for the formation of heme is deficient; it can manifest as abdominal pain, weakness, confusion, delirium, psychotic episodes, seizures, and erosion of the skin due to sunlight sensitivity. In some cases, excess porphyrins are eliminated in the urine, imparting a reddish-purple color that turns dark on exposure to light.

In the 1960s, mother and son psychiatrists Ida Macalpine and Richard Hunter proposed a theory in the *British Medical Journal* that George III was afflicted with porphyria. Their posthumous diagnosis was based on contemporary reports by the king's physicians saved in the Royal Archives and the British Library. These document George's abdominal pains, muscle weakness, hoarse voice, and periodic bizarre behavior, described by one of his doctors as "alienation of mind." The "final proof," Macalpine and Hunter concluded, was in reports that the King passed discolored urine, blue in one particular case.

The allegation that the king's occasional extreme changes in mood, eccentric behavior, and delusions were caused by a physical disease captured the imagination of the media and spurred numerous articles in the popular press. There were even suggestions that the loss of the American colonies was due to a mentally disturbed king making poor military decisions. That seems far-fetched since the first serious episode of psychotic illness was recorded in 1788, well after the Revolutionary War.

The porphyria hypothesis has been challenged by a number of scientists who have examined the "evidence" provided by Macalpine and Hunter and have concluded that it doesn't stand up to scrutiny. The abdominal pains and muscle weakness as described are not consistent with what is normally seen in porphyria, but it is the supposed discoloration of the urine that particularly raised the ire of the critics. The records do show a few reports of "bloody water," one of "bilious urine," the meaning of which is obscure, and one case of urine tinged blue. These, particularly the blue color, are hardly evidence of the purple urine that is associated with porphyria. One researcher has proposed that the blue may have been due to an extract of the gentian flower that was given to the king as a remedy. In any case, the proponents of the porphyria theory ignored the numerous reports about the king passing normal colored urine during his psychotic episodes and cherry-picked the few cases that they then stretched to fit their theory. Some critics claim that Macalpine and Hunter distorted the evidence to promote their philosophical agenda that mental illness was not an entity by itself but was a manifestation of some metabolic disorder.

How then do these critics explain King George's episodes of peculiar behavior? They make a strong case for his suffering from bipolar disorder, which is consistent with the periodic appearance of the symptoms and has been further buttressed by a fascinating paper published in 2017 that examined the digitized version of a multitude of letters written by George III over his long reign. Computational approaches were used to compare syntax, stylistic patterns, and ideation with material written by people diagnosed with bipolar disease. The conclusion was that George's letters composed during his periods of mania reflected changes that were consistent with those seen in the writings of bipolar patients.

While the connection of King George III to porphyria is certainly tenuous, the association of the disease with the legend of the vampire is even more speculative. The curtain on this scenario was raised by noted Canadian biochemist Dr. David Dolphin of the University of British Columbia. At a conference in 1985, he conjectured about how victims of porphyria, with their symptoms of sensitivity to light, mottled skin, and reddish teeth due to porphyrin buildup, may have created the legend of the vampire. Dolphin is a highly respected chemist with a long history of specializing in porphyrin research who has developed a widely used porphyrin-based medication for macular degeneration. But his 1985 talk about vampires was somewhat whimsical, designed to highlight porphyrin research, and was never published in the scientific literature. However, reporters at the conference snapped up the theory and came up with clever headlines like "Chemist Goes to Bat for Vampires," implying that there may be more to vampires than a myth. There isn't.

While such musings may be entertaining, there is a serious side to the alleged porphyrin-vampire connection. After articles about vampires and porphyria flooded the media, one physician reported a patient with the disease who became depressed and needed reassurance that he was not descended from vampires and would not turn into one. It is time to drive a wooden stake through the heart of any speculation that victims of porphyria are really vampires.

THE MYSTERIOUS PINEAL GLAND

I was confused. Just about every discussion of the pineal gland, be it in a scientific publication or a lay article, attributes the name of the tiny, rice-grain-sized gland found in the geometric center of the brain to its shape that is claimed to resemble a pinecone. Reference is usually made to Galen, the ancient Greek physician who is said to have first described the distinctive cone shape of the gland. That's pretty interesting I thought, but no amount of searching through anatomical texts or online images revealed a picture of the gland that looked anything like a pinecone. Time to go back to the supposed original reference! It turns out that Galen never likened the shape to a pinecone; he thought the gland resembled a pine nut! All those articles that describe the pineal gland as being shaped like a pinecone are wrong. Not an earth-shaking error, but an example of how a mistake can be perpetuated and becomes "true" by repetition.

There is something else to note about Galen's description of the pineal gland. He never carried out a dissection of a human brain! Galen spent most of his life in Rome, and Roman laws prohibited the dissection of human bodies. That was a problem because Galen believed that knowledge could be gained not through philosophical meanderings, but through experimentation. "I am a man," he said, "who attends only to what can be perceived by the senses, recognizing nothing except that which can be ascertained by the senses alone with the help of observation." So, with an insatiable search for knowledge, he turned to animals and dissected monkeys, cats, dogs, weasels, camels, lions, wolves, stags, bears, mice, and even one elephant. In one lecture to medical students entitled "On the Brain," he describes a systematic dissection of the brain of an ox, and it is here that he identifies the pineal gland and likens it to a pine nut. Of course, he would have had no way of knowing the function of this gland.

We don't hear much of the pineal gland until the seventeenth century when French philosopher, mathematician, and scientist Rene Descartes introduced the notion that the pineal was the connection between the body and the soul and was the place in which all our thoughts are formed. How he came to this conclusion isn't clear, but he seems to have been taken by the location of the gland, essentially in the center of the brain, and by the fact that unlike many structures that are duplicated in the two halves of the brain, the pineal gland is unpaired.

Science finally entered the scene in 1917 with the publication of an experiment by Carey Pratt McCord and Floyd Allen in which they fed newly hatched tadpoles with an extract of bovine pineal glands. Why such a seemingly strange experiment? Many types of frogs change color depending on such factors as temperature, exposure to light, need for camouflage, and even anxiety. The researchers were attempting to explore triggers for such color changes and tried feeding different lots of tadpoles with the likes of breadcrumbs, hemp seeds, desiccated retinae from beef eyes, and extracts of cow pineal glands. Only the latter produced a change. And what a change! The tadpoles became virtually translucent as the pineal extract somehow interfered with the production of the dark pigment, melanin.

Over the next forty years, various investigators reported that injection of pineal gland extracts into tadpoles, frogs, toads, and fish produced a lightening of skin color. This was attributed to interference with melanocyte stimulating hormone (MSH) produced by skin cells that stimulates the production of melanin, the pigment that is a major determinant of skin color. In 1958, researchers from Yale University led by Aaron Lerner isolated a compound from beef pineals that was responsible for the interference with MSH and suggested that it be called *melatonin*.

Once it became clear that melatonin had a biological function, interest in the compound increased especially after 1975 when Richard Wurtman at MIT found that melatonin content of urine samples from healthy adult volunteers collected between 11 p.m. and 7 a.m. was higher than in daytime samples. Melatonin production, it seemed, varied with the sleep-wake cycle. That brought up the question of whether melatonin could be used to regulate a cycle that is out of whack, such as in people who experience jet lag or have trouble falling asleep. Dr. Wurtman explored this possibility with twenty volunteers who were given either synthetic melatonin or a placebo. Subjects who got melatonin fell asleep significantly faster than those who received the placebo and also had increased duration of sleep. Based on these results, Wurtman filed a patent for using from 0.3 to 1 mg of melatonin, the doses used in his study, to help with sleep problems.

However, in the opinion of the U.S. FDA, since melatonin was known to occur naturally in some foods such as tart cherries, nuts, and goji berries, it fell into the category of a dietary supplement and could be sold without a prescription. Given that many people struggle with insomnia, it was no surprise that sales of melatonin supplements boomed. "More is better" became a common promotional theme and pills containing up to 10 milligrams flooded the market. Wurtman warned that the safety and efficacy of high doses had not been established, but "if a little is good, more must be better" held sway with the public. There was yet another issue. Dietary supplements are not regulated as carefully as prescription drugs, and surveys found that melatonin content ranged from being 83 percent less to 478 percent more than declared on the label!

Since Wurtman's original study, many melatonin trials have been carried out, but results are inconsistent. My own anecdotal evidence seems to be in step with those findings. I find that a sublingual spray works better than a pill. A single spray, that according to the label contains 1 mg melatonin used one to two hours before the intended bedtime works. At least sometimes. More is definitely not better.

Something else to keep in mind. Light impairs the production of melatonin. Especially blue wavelengths emitted by TV screens, laptops, and even alarm clocks. Watching TV in bed while using an iPad to check the latest research on melatonin is not conducive to good sleep.

Since melatonin for me only works sometimes, and since I'm addicted to my screens, I was intrigued by headlines about a study in which subjects who slept under weighted blankets had higher levels of melatonin in their saliva. Alas, as with the pinecone story, I'm also addicted to checking the facts. It turns out that while subjects with the weighted blankets had higher melatonin levels, their total sleep time did not improve. Maybe a glass of milk would work. Turns out it contains melatonin.

FAR INFRARED FAR FROM SCIENCE

What gets between Tom Brady and his pajamas? Bioceramic microparticles! Truth be told, the tiny crystals of silicon dioxide and titanium dioxide aren't exactly between Tom and his pajamas, they are embedded in the polyester fibers. Why? They are said to be the key to helping sore muscles recover and to facilitating better sleep. Who says so? Tom Brady, who without a doubt is one of the best quarterbacks in NFL history. However, completing passes on the gridiron is not the same as completing and passing courses on science.

In a promotional video for the Under Armour Company, Tom opines that "without the sleepwear, I don't really feel like I would have been able to achieve the things that I have done and hope to continue to do." That sleepwear is made of Celliant, a proprietary textile that claims to capture body heat and reflect its far infrared (FIR) component back into the body. This has the effect of "increasing local circulation and cellular oxygenation resulting in more energy, endurance, strength, stamina, comfort, quicker recovery, and better sleep." "Tested and proven through clinical trials" claims Hologenix, the manufacturer of the fabric that Brady so enthusiastically endorses. It should be remembered, though, that Tom was winning Super Bowls long before he ever heard of bioceramics.

"Tested and proven through clinical trials" is indeed an alluring claim, but it doesn't exactly mean that the product or procedure being evaluated was proven to be of practical value. For example, one of the studies quoted was published with the title: "Effect of Celliant Armbands on Grip Strength in Subjects with Chronic Wrist and Elbow Pain: Randomized Double-Blind Placebo-Controlled Trial." The researchers found that grip strength as measured by a device known as a dynamometer increased by about 9 percent more in subjects who wore the ceramic-impregnated armband than in those who wore the placebo. Statistically this was unlikely to be due to chance and therefore the armband was judged to be "effective." But no measurements of any effect on pain were carried out, so this study doesn't really provide any thunderous support for the benefits of the Celliant fiber.

How about a study that seems to be more relevant to the pajama effect? Let's consider this one: "Randomized Controlled Trial Comparing the Effects of Far Infrared Emitting Ceramic Fabric Shirts and Control Polyester Shirts on Transcutaneous Oxygen Pressure." In this case, the level of oxygen circulating just below the skin was measured, and since blood flow is important in wound healing, transcutaneous oxygen measurement can be used to gauge the ability of tissue to effectively heal. After wearing the ceramic fabric shirt for ninety minutes, the transcutaneous oxygen pressure was about 7 percent higher than after wearing the placebo shirts for the same time period. An interesting finding, but that is a very small difference, and no measurements were made of muscle recovery, stamina, or sleep. Not much to hang a hat on.

Not included in the list of "clinical evidence" studies on Hologenix's website is one published in 2021: "Utilisation of Far Infrared-Emitting Garments for Optimizing Performance and Recovery in Sport: Real Potential or New Fad? A Systematic Review." Researchers scrutinized the scientific literature for trials that involved the wearing of far infrared emitting garments and measurements of any effects on performance or physiology and found eleven trials that met the criteria of having been well carried out. None showed a statistically significant effect on performance. Two suggested a slight reduction in muscle soreness and one showed a small effect on sleep duration. Not exactly compelling evidence.

But let's not throw far infrared emitting technology under the bus just yet. Let's let mice, frogs, and rabbits have their say. In one study, mouse muscle cells were cultured with ceramic powder under the culture dishes and showed less signs of oxidative stress and elevated levels of nitric oxide, a neurotransmitter that plays a role in vasodilation and enhanced blood flow. In frog skeletal muscles, far infrared emitting ceramics delayed the onset of fatigue induced by muscle contractions. And when arthritis was chemically induced in rabbits, those in cages surrounded by paper sheets impregnated with a thin layer of ceramic powder experienced greater reduction in inflammation than a control group surrounded by the same sheets without the powder. But of course, we are not mice, frogs, or rabbits.

More compelling evidence can be found in studies that have examined far infrared rays (FIR) emitted from electrically powered devices such as infrared lamps and saunas. Wounds in mice heal faster when exposed to FIR, and when mice with experimentally reduced circulation in a hind leg are placed in a far infrared sauna, they grow new blood vessels, a process known as angiogenesis. Again, these are mice. But there is some human evidence as well. In Japan, FIR saunas are extensively used for treatment of cardiovascular conditions and have been shown to improve cardiac output.

Then we have claims like "the healing power of the body is now known to be literally limitless when quantum coherence is restored to the biofield." Of course, it is their "Photon Genius Energy Infrared Sauna" that can restore such coherence. Incoherent nonsense. So is any claim that infrared saunas detoxify the body. The liver, kidneys, and digestive tract do that.

Now back to Tom Brady's pajamas. At $200, they are not cheap. But given that at age forty-five Tom can still perform like a young buck, and that he attributes some of the success to his inflammation-reducing pajamas (albeit without any evidence), it is not surprising that many Tom wannabees have forked out the money. They have also blogged extensively about their experience. The PJs are comfortable and not overly warm, they say. But as far as any restorative effect goes? Not noticeable. Maybe they should try wearing them in a far infrared sauna.

By the way, Tom follows quite a healthy diet, works out a great deal, and doesn't use any electronic screens before going to bed, which he does at 8:30. No wonder he sleeps well. And we know how important sleep is for athletic, or indeed, any performance. I suspect Tom would sleep just as well without the bioceramics. Well, maybe a bit better with them, given what he is paid for their endorsement.

CAN YOU EAT TO BEAT DISEASE?

Just about any publication that explores the role of diet in disease invokes Hippocrates's famous dictum, "Let food be thy medicine, and let medicine be thy food." Actually, there is no record of the famous ancient Greek physician ever having said this, although it is clear from the writings of the Hippocratic authors that Greeks did believe that disease and food were linked. "Hippocratic authors" is the correct terminology because historians concur that the works attributed to Hippocrates are compilations of his own writings and those of a number of his followers.

It isn't surprising that the Greek physicians associated food with disease because obviously without food there is no life. However, whatever dietary interventions were recommended were not based on any evidence but rather on the belief that the body is constituted of the four humors, namely blood, phlegm, yellow bile, and black bile. Illness was due to an imbalance in these humors and was to be treated with foods that would restore balance. For example, pain in the joints was described as being due to blood being contaminated and coagulated by phlegm and bile. The treatment was barley water. In terms of science this makes no sense, but undoubtedly some patients found relief through the power of the placebo.

We have come a long way from ancient Greek medicine, and today our knowledge about nutrition is quite extensive especially when it comes to prevention of disease. The role of vitamins, minerals, proteins, fats, and carbohydrates in the diet is well established, and we are also getting a handle on how some naturally occurring compounds in plants may contribute to the maintenance of health. However, the science of treating disease with food is much more nebulous.

I first became acquainted with Dr. William Li's exploits when I was looking into the work of Dr. Judah Folkman, the recognized pioneer of anti-angiogenesis research. *Angiogenesis* is the term used to describe the growth of blood vessels, and consequently anti-angiogenesis is the impairment of blood vessel growth. Tumors need nutrients to grow, and they generate blood vessels through which these are supplied. In theory then, any factor that has anti-angiogenesis effects can rob a tumor of nutrients and cause it to shrink. Dr. Li trained under Folkman and went on to found the Anti-Angiogenesis Society that fosters research aimed at restricting tumor growth. He is a very sound researcher and has published widely about angiogenesis in top journals.

It was in 2019 that Dr. Li ventured into the area of nutrition with his book *Eat to Beat Disease*. It has been a major bestseller and his TED Talk on the same topic has been viewed by millions. He is undoubtedly an "influencer," but the question is to what extent can one benefit from his influence. Dr. Li is far from being a master of woo as some have claimed, but some of his conclusions about the relationship between diet and disease should be marked with an asterisk.

He describes, correctly, that health depends on a number of factors that include a properly functioning immune system, diversity of gut bacteria, an ability to eliminate cancer stem cells, effective repair of damaged DNA, and impairment of the growth of blood vessels that feed tumors while enhancing the growth of "good" blood vessels such as the ones that feed the heart. Dr. Li provides ample literature references of how nutrients in food can affect these determinants of health.

These include "bioactives" in plums, apples, chicken soup, and olive oil to curb inflammation. Anti-angiogenesis factors are to be found in salmon, tomatoes, onions, and blueberries. Purple sweet potatoes battle cancer stem cells while compounds in oysters and turmeric are said to repair errors in DNA. Immune system activators are to be found in broccoli sprouts, peppers, garlic, liquorice root, and mushrooms. Eggplant encourages the growth of "good" blood vessels. Walnuts, yogurt, sauerkraut, beans, kiwis, and cocoa encourage the growth of "healthy" bacteria in the gut. The problem, though, is that the studies from which all this information is drawn come from laboratory studies using cells or animals. Extrapolating these studies to humans is not totally unreasonable but is a touch too cavalier.

There is absolutely no doubt that along with exercise and favorable genetics, diet is a key preventative when it comes to disease. But curing disease by eating specific foods is another matter. The body is not a large test tube and humans are not giant mice. A substance that has an anti-angiogenesis effect in the laboratory cannot be assumed to have a similar effect in a cancer patient. Indeed, there is also the issue of placing an extra burden on a patient who may believe that their cancer isn't "being beaten" by the foods they are eating because they have not properly adhered to the recommendations. That being said, there is nothing harmful about Dr. Li's recommendations. Indeed, they are pretty well in step with what most researchers recommend, namely a diet based on fruits, vegetables, whole grains, nuts, extra-virgin olive oil, and fish, with limited meats and processed foods. Indeed, an analysis of nutritional studies reported in 2018 in the *Journal of the American Medical Association* estimated that just ten foods account for nearly half of all U.S. deaths from heart disease, stroke, and type 2 diabetes. These deaths occur due to people eating too few nuts, seeds, seafood omega-3, vegetables, fruits, and grains, or eating too much sodium, processed meat, unprocessed red meat, and sugary beverages.

I think Dr. Li overplays the *Eat to Beat Disease* card, but his views about anti-angiogenesis and nutrition are sound. While his opinion about treatment of various ailments with specific foods falls into the speculative category, his dietary recommendations in terms of prevention of disease are on a firm footing. Given the rather sad state of the average Western diet, most people would benefit from paying heed to Dr. Li's nutritional advice. *Eat to Beat Disease* is worth reading, but *Eat to Prevent Disease* would have been a more realistic title.

TIME-RESTRICTED EATING . . . OR NOT . . .

Nutritional research publications can be maddening. Their sheer volume is overwhelming, but once the poorly done studies are filtered out, once you eliminate the ones that may have some statistical but no clinical significance, get rid of the ones that make grandiose claims about health effects based on food questionnaires administered on a single occasion, purge the ones that over-interpret results based on too few subjects, exclude those that imply long-term effects based on short-term trials, and disregard studies that unrealistically extrapolate animal data to people, you don't have much left to chew on. But let's nibble away at the studies that explore a current hot topic in "wellness," which is time-restricted eating. The proposition is that when it comes to health status, it is not only what we eat that is important, but also when we eat.

While there certainly is controversy about details, there is a general consensus about the "what." A diet should be based mostly on vegetables, fruits, and whole grains, with minimal processed foods, especially ones high in sugar and salt. Fish is preferred over other flesh foods, and extra-virgin olive oil over refined seed oils. Saturated fats, such as in butter and in red meat should be limited, as well as foods charred by high heat. Alcohol no more often than a couple of times a week, and soft drinks as close to never as possible. With that out of the way, let's get down to the "when."

It was back in 1935 that Cornell University nutritionist Dr. Clive McCay discovered that mice fed a diet that reduced calorie intake by

30 percent were physically more active and far less prone to diseases of advanced age than their free-grazing laboratory mates. Furthermore, they lived about 40 percent longer! At the time, the interpretation was that the benefits were due to a reduced production of tissue-damaging reactive oxygen species (ROS) that are byproducts of the reaction between glucose and oxygen, the reaction that produces the energy needed to fuel the life processes in cells. Chemically speaking, ROS are free radicals, electron-deficient species that are ready to steal electrons from any molecule they may encounter. Since electrons are the glue that bind atoms together in molecules, their loss results in bond breakage. Should the molecules affected be proteins or nucleic acids, the consequence can be disease or accelerated aging. But if there is less food to metabolize, goes the argument, fewer free radicals are produced with the result being enhanced longevity.

Since McCay's original observation, numerous studies using fruit flies, mice, dogs, and monkeys have demonstrated that caloric restriction increases longevity. However, reduced free radical formation may not be the only factor involved. Not recognized in the early rodent experiments was the fact that the animals consumed their restricted food allotment within a few hours of it being provided, meaning that they had long periods of fasting.

In a normal state, cells "burn" glucose, provided by the diet, to produce the energy needed to sustain life. In a fasting state, with no glucose being provided, a backup system is engaged. Cells, instead of "burning" glucose, switch to metabolizing stored fats. This involves the breakdown of fats to yield acetone, acetoacetate, and beta-hydroxybutyrate, the so-called ketone bodies that serve as an alternate fuel for energy production. But ketosis, as this fat-burning process is called, is also a signal to the body that there is no food coming in, a crisis situation. The metabolic switch to burning ketones instead of glucose then triggers a number of cellular responses aimed at survival. Cells begin to crank out various molecules that repair DNA, reduce inflammation, regulate glucose sensitivity, and break down damaged cells (autophagy). All these processes can benefit health.

This brings up the question of whether the benefits of a calorie-restricted diet seen in animals may be a function not only of the reduced calorie content, but also of the time frame during which no food is consumed. Is there an optimal way, researchers wondered, to incorporate fasting into a dietary regimen? What if instead of just cutting down on calories, attention were paid to when the meals that make up that restricted calorie diet are eaten? Thus was born the concept of time-restricted eating, or its alternative designation, intermittent fasting.

Several regimens have been proposed. Eating a regular diet on five days and cutting calories down to 500 to 700 on two days a week (5:2 fast), doing the same on alternate days of the week (4:3 fast), or fasting for fourteen to sixteen hours a day (daily time-restricted eating) have all been tried. In the latter case, no restrictions are placed on calories during the eight to ten hours when food is consumed, but experiments have shown that this automatically results in a reduction of calories because night-time snacking is eliminated. While most of the trials involving these regimens have resulted in weight loss, the benefits such as improvement in glucose regulation, blood pressure, inflammation, and loss of abdominal fat go beyond what would be expected for weight reduction. For example, in one study, women were assigned either to a 5:2 intermittent fasting regimen or a daily 25 percent reduced calorie diet. Over six months, both groups lost the same amount of weight, but the 5:2 group had improved insulin sensitivity and a larger reduction in waist circumference.

Other studies involving intermittent fasting have shown better running endurance, improvements in the HDL/LDL cholesterol ratio, reduction in free radical activity, and reduced markers of systemic inflammation. Some preliminary studies have also shown suppressed tumor growth in a number of cancers. In animal models, alternate day fasting can delay the onset and progression of neurological diseases such as Alzheimer's,

Parkinson's, and multiple sclerosis. There are even suggestions that intermittent fasting can improve memory and cognitive performance.

The evidence of benefits continues to accumulate. In a widely quoted study, one group of mice was given access to food only during a nine-hour period, while those in a control group were able to eat whenever they liked. The two groups actually ended up eating roughly the same amount of food, so at least in this case, whatever results were obtained could not be ascribed to a difference in caloric intake. After seven weeks, tissue samples were taken from multiple organs and examined for any changes in gene expression. Genes code for the production of proteins, so basically the researchers measured whether the production of various proteins increased or decreased. Genes that code for proteins responsible for inflammation were found to be less active, while genes that produce proteins that repair damage to DNA and ones that inhibit cancer cell survival geared up. But, of course, mice are not men or women.

So, what about men or women? One interesting study examined changes in a number of proteins produced as a result of eating only during a ten-hour period and fasting for fourteen hours. The subjects, eight men and six women, were all observers of the Muslim religious month of Ramadan during which no food or drink is consumed between dawn and sunset. They were specifically selected because each one met at least three criteria of metabolic syndrome, defined as central obesity, insulin resistance, high blood pressure, high levels of triglycerides, and high cholesterol. These parameters are easily monitored and can provide information about the health effects of fasting in addition to changes in gene expression.

All of the markers of metabolic syndrome shifted in the right direction during the month of the fourteen-hour fast, as did proteins involved in destroying cancer cells, repairing DNA, and improving immune function. All very interesting, but the experimental group was small, and the study period of a month was short. Also, the subjects all had metabolic syndrome, and calorie intake was not considered. Basically, not much can be inferred as far as the general population goes.

That though is not the case for a study that compared the effects of eating an early or late dinner on glucose levels, insulin production, triglyceride levels and fatty acid oxidation which is a measure of ketosis. Subjects ate their dinner either at 6 or 10 p.m. and then had their blood chemistry monitored every hour through an intravenous line. The late dinner resulted in greater glucose intolerance and reduced fatty acid oxidation, both of which can promote obesity. Why should this happen? During sleep, metabolism normally winds down since the body needs less energy. Therefore, ingested glucose and fats are not burned for energy, but rather end up being stored as fat. If dinner is eaten earlier, metabolism remains active until sleep time and less fat ends up being stored. This study would seem to corroborate the benefits of the daily time-restricted fast since if no food is eaten after late afternoon, the reduced metabolism associated with sleep is less of an issue because most of the food will have been metabolized in the five or six hours between the last meal and sleep.

Now, just as I was ready to wrap things up with a final praise of intermittent fasting schemes, I learned of two recently published papers in respected journals. One found that in adults over the age of forty, a time interval of fewer than 4.5 hours between meals, which essentially means time-restricted eating, was associated with earlier death! Yikes! The second study asked participants to use an app to record the timing of their meals and then went on to relate this to their body weight as documented in their medical records over a ten-year period. Weight changes were not associated with the time between the first and last meals, which would seem to argue against trying to lose weight by time-restricted eating.

Where does all this leave us? As is the case with almost every aspect of nutrition, there is controversy, and studies can be found to back up each side. Separating the wheat from the chaff is challenging and requires an extensive review of studies to try to get a handle on the preponderance of evidence. At this point, that evidence indicates caloric restriction to be a factor in reducing markers of disease and longevity, but to make recommendations, especially ones that are difficult to institute, we need more than markers. We need long-term human trials, with a significant number of subjects that compare regular diets, reduced calorie diets, and intermittent fasting with end points of disease or death. Such lengthy trials are difficult if not impossible to finance, organize, and monitor. In their absence, we are reduced to making educated guesses.

Since none of the calorie-restricted regimens has shown any risk, there seems to be no harm in giving one or another a shot, whether it be for weight loss or just enhanced health and perhaps a longer life. But I suspect most people would not be able to endure calorie restriction over the long term. There is just too much pleasure to be had from eating. However, having an early dinner and then fasting until bedtime may be a challenge that can be met and may be worth a try. At least until the next study comes out telling us that life expectancy in Spain, where dinners are traditionally eaten late at night, is longer than in North America.

Obviously, the field of nutritional research is very fertile and there are many plants to harvest, but we do have to watch out for the weeds.

EVERYONE IS TALKING ABOUT OZEMPIC

Glucagon-like peptide-1 agonist. Sounds like a mouthful, but such substances may actually keep you from filling your mouth. Ozempic, Wegovy, and Mounjaro, the GLP-1 agonists that have been basking in the spotlight, may be the long-awaited medications that can help win the battle against obesity. "May," though, is an important qualifier.

The use of GLP-1 agonists for weight control is often described somewhat overenthusiastically as a "breakthrough" or "giant leap forward." However, history teaches us that science rarely progresses by giant leaps; discoveries are the result of a series of small steps. Still, the story of GLP-1 agonists does indeed begin with a giant leap, one that was taken in 1902 by physiologists Ernest Starling and William Bayliss at London's University College.

"On January 16th, 1902, a bitch of about 6 kilos weight . . ." is the epic beginning to Starling and Bayliss's description of their "critical experiment" in the September issue of that year's *Journal of Physiology*. That experiment involved severing all nerves around the pancreas and the duodenum, the beginning of the small intestine, of the female dog in question, and then injecting a small amount of acid, such as produced by the stomach during digestion, into the duodenum. Remarkably, this stimulated secretion of pancreatic juices despite there being no connection via nerves! Perhaps even more remarkable was the observation that intravenous injection of an extract of the intestinal mucosa mimicked the action of the acid. Clearly, some chemical secreted in the intestines was able to send a message to the pancreas through the bloodstream.

Starling and Bayliss named this substance *secretin* and coined the term *hormone*, from the Greek for "stir into action," to describe substances produced in one part of the body capable of stimulating action elsewhere by traveling through the bloodstream. This was the beginning of the field of endocrinology, from the Greek terms *endo*, meaning within, and *krine*, meaning to secrete.

Starling and Bayliss's seminal paper stimulated further research into duodenal secretions, and by the 1970s a number of details emerged. Introducing glucose into the intestine sent a message to the pancreas to release insulin into the bloodstream. That message was in the form of two hormones, glucose-dependent insulinotropic polypeptide (GIP) and glucagon-like peptide-1 (GLP-1). An obvious corollary to this discovery was the potential of these hormones to treat diabetes. In a 1987 landmark paper, Dr. Daniel Drucker of the University of Toronto described GLP-1 as being composed of a chain of thirty amino acids formed in the intestine but went on to explain that the potential of using this peptide as a treatment presented a big challenge. Although by the 1980s the sequence of amino acids in such a peptide could be readily determined, and a laboratory synthesis accomplished, there was a major problem. GLP-1 had a very short half-life in the blood, meaning that a diabetic patient would have to inject it every few minutes. This was obviously not practical, and the search was on to find a way to modify the molecular structure of GLP-1 with an aim of retaining its insulin boosting effect while extending its survival in the blood.

It is at this point that science got a monstrous boost from an unlikely source, a foot-long lizard found mainly in Arizona and New Mexico. The Gila monster eats only once or twice a year, a phenomenon that interested Dr. Jean-Pierre Raufman, a gastroenterologist at the National Institutes of Health. He found that the lizard's saliva contains biologically active molecules that cause enlargement of the pancreas in test animals. Since the pancreas produces insulin, Raufman's discovery intrigued Dr. John Eng, an endocrinologist at the Veterans Administration Medical Center in New York. Eng was well equipped to explore this further, having trained under Rosalyn Yalow, recipient of the 1977 Nobel Prize in Physiology or Medicine for the development of radioimmunoassay, a technique that allows for the measurement of tiny amounts of biological substances in blood, insulin being a typical example.

In 1992, following up on Raufman's work, Eng discovered that a compound in the lizard's saliva, which he named *exendin-4*, was a peptide that had an amino acid sequence similar to that of GLP-1. Could this perhaps be of some use in the treatment of diabetes? he wondered. Indeed, exendin-4 stimulated insulin release and, importantly, it did not get degraded as quickly in the blood as GLP-1! Dr. Eng patented exendin-4 as a drug and in 1996 licensed it to Amylin, a pharmaceutical company that then partnered with Eli Lilly to introduce exenatide, a synthetic version of exendin-4, as Byetta for the treatment of type 2 diabetes. A twice-daily injection stimulated insulin secretion whenever glucose entered the intestines, and unlike drugs that boost insulin secretion indiscriminately, had a reduced risk of causing hypoglycemia. And there was an added benefit. The drug reduced appetite and resulted in weight loss!

Parallel to the development of exenatide, researchers at Novo Nordisk, a Danish pharmaceutical company, were also tackling the problem of GLP-1 breakdown. Their idea was to link the peptide to another molecule that could ferry it around the bloodstream and prevent contact with the enzymes that normally degrade it. Extensive trials resulted in exchanging a few amino acids for others in GLP-1 and joining the modified version to human serum albumin, a protein naturally produced in the liver. The link was formulated with an array of glutamate and fatty acid molecules in a proprietary fashion. This GLP-1 agonist, so-called because of its fit into receptors on pancreatic cells just like natural GLP-1, went on the market as liraglutide, a daily injection for the treatment of type 2 diabetes. It was subsequently approved for obesity as well. Further subtle adjustments in molecular structure led to semaglutide, which required only a weekly injection. It was approved for diabetes as Ozempic, and at a slightly higher dose for obesity as Wegovy. Mounjaro, the Eli Lilly Company's brand name for tirzepatide is the newest kid on the block and promises to be even more effective at weight control since it mimics not only GLP-1 but GIP as well.

There are some caveats to be considered before anointing these drugs as the Holy Grail of weight control. At around $1000 a month, they are expensive, and use has to be continuous because weight is regained if they are stopped. Long-term effects, especially on the pancreas and the thyroid gland, are unknown for the simple reason that a long term has not passed since the drug was introduced. Nausea and vomiting are not infrequent side effects and can be serious. Curiously, there have also been reports of bizarre dreams. In one case, an Ozempic user described playing baseball in front of a hostile crowd and being rescued by Oprah Winfrey in a go-kart. On the positive side, some patients on semaglutide who also struggle with addiction have reported a reduced desire for tobacco and alcohol. According to a study by the manufacturer that has not yet been replicated, Wegovy was shown to reduce the risk of heart attacks and strokes in patients with cardiovascular disease by 20 percent. Whether this is due to a loss in weight or some other effect that the drug has on the cardiovascular system isn't clear.

The high price of Ozempic generated a copycat industry staffed by compounding pharmacists who claim to be able to make the drug in their lab and provide it at a fraction of the price of the brand-name drug. Compounding is a legitimate segment of the pharmacy business in which pharmacists make special versions of drugs for people who may have allergies to an ingredient in the standard version, or perhaps require a liquid preparation instead of a pill. However, in the case of semaglutide, it is hard to know what they are actually compounding, since Novo Nordisk does not sell the active ingredient to outsiders and has not published details of the manufacturing process. Semaglutide itself can be synthesized and is indeed produced by a number of companies, mostly in China. But the technology Novo Nordisk used to link the molecule to albumin, which is critical for its efficacy, is proprietary. It is unlikely that cheaper versions of semaglutide made by compounding pharmacies, or sold online, are equivalent to Ozempic.

The race for an effective weight control drug is understandable since obesity is a risk factor for diabetes, cardiovascular disease, and cancer. However, in the case of GLP-1 agonists for weight loss, the foot on the gas pedal may be a touch too heavy. Not all segments of the population have been represented in clinical trials, and studies have not yet shown improved health outcomes in users, although one trial did show reductions in blood pressure and cholesterol that paralleled weight loss.

Physicians have to carefully weigh whether potential benefits for an overweight person outweigh the risks. While it is likely that someone with a body mass index over twenty-seven and having at least one risk factor for cardiovascular disease will benefit, the same cannot be assumed for someone who may be packing an extra ten to twenty pounds without any risk factors. GLP-1 analogues are not for people trying to squeeze into last year's bathing suit. Where a squeeze is needed is on the overly enthusiastic promotion of Ozempic as a silver bullet to deflate the obesity epidemic.

It comes as no surprise that as more and more people are using weight loss drugs such as Ozempic, Wegovy, and Mounjaro, there are more reports of adverse reactions. There are a couple of reasons for this. First, at most a few thousand subjects are studied in Phase 3 trials, the ones that determine if a drug is safe and effective enough for marketing. This means that a problem that can affect, let's say, one in ten thousand people is not likely to be picked up. Yet, the numbers affected can be very large when hundreds of thousands of people take the medication. There is also the issue that Phase 3 trials last at most three to four years, so long-term effects can also escape detection.

The second possibility for increased complaints is that any condition is more likely to be noted in a large group of people than in a small group. Quite naturally, anyone who experiences a health problem wonders about what the cause may have been. If a new medication was recently started, it may be suspected as the cause. It is important to constantly keep in mind that correlation is not the same as causation.

That being said, there are enough reports of side effects such as nausea, vomiting, constipation, diarrhea, and stomach pains to conclude that there is a cause-and-effect relationship with GLP-1 agonists. And there is now concern about another effect that is being reported, gastroparesis, which means stomach paralysis. The incidence is not clear, but one woman in the U.S. has launched a lawsuit against the manufacturers claiming that her stomach was paralyzed to such an extent that it would not pass food into her intestines. Whenever she ate, the food would come right back up! The vomiting was so severe, she says, that her teeth started to fall out. She claims that while possible side effects are listed on the informational insert that comes with GLP-1 agonists, she was not made aware of the possibility that her stomach would be paralyzed. It will be interesting to see how this case comes out, but in the litigious U.S., class action lawsuits are on the horizon.

Anesthesiologists have encountered another issue with these drugs. It is standard procedure to have patients fast when they are being readied for surgery or other procedures that require sedation to ensure their stomachs are empty. If there is food in the stomach, there is a chance that under sedation it can be aspirated into the lungs. However, GLP-1 agonists reduce the rate at which the stomach is emptied, so the patients on these drugs need to fast longer or stop taking the medication. Yet another concern that has surfaced is the possibility that suicidal ideation may be linked to the use of these weight control drugs. If this turns out to be a legitimate link, it is undoubtedly rare, and one that would not have been picked up in the studies that led to approval.

Ozempic, Wegovy, and Mounjaro are turning out to be some of the most successful drugs ever churned out by pharmaceutical companies. That's because when it comes to treating type 2 diabetes and causing weight loss, they really do work. Sales are set to soar to even greater heights because researchers are discovering that these drugs which mimic the action of certain hormones produced in the body may produce other benefits as well. Reductions in the risk of heart, kidney, and liver disease are being observed, as well as reduced blood pressure and arthritic pain. Add to this a decline in addictive behaviors such as heavy alcohol drinking, opioid use, and gambling, and a reduction in the risk of Alzheimer's disease. No surprise then that the pharmaceutical industry is driven to come up with even better versions of these drugs.

Novo Nordisk, the company that produces Ozempic and Wegovy, is hoping that its experimental drug amycretin will be the next blockbuster. While semaglutide, the active ingredient in Ozempic and Wegovy, mimics the action of GLP-1, a hormone produced in the intestine that regulates appetite and feelings of fullness, amycretin ups the ante. Besides mimicking GLP-1, it also activates receptors for amylin, a hormone that is produced in the pancreas and is involved in suppressing appetite and reducing food intake. In a preliminary trial, patients on amycretin lost 13 percent of their body weight after twelve weeks, double what has been seen with Wegovy. What makes this drug even more appealing is that, unlike Ozempic, Wegovy, and Mounjaro, instead of being injected, it can be administered as a pill taken once a day. Another potential benefit is a reduced risk of such gastrointestinal symptoms as nausea, vomiting, and abdominal pain usually seen with GLP-1 agonists.

The Eli Lilly Company believes it may have a winner in its experimental drug retatrutide, which in addition to being a GLP-1 agonist mimics the action of two other hormones that control appetite.

Retatrutide has to be injected, but Lilly is also experimenting with orforglipron, yet another GLP-1 agonist. The advantage with this drug is that not only can it be taken orally, it is also much easier to produce. Unlike the other drugs in this class that are peptides, that is chains of amino acids, orforglipron is a small molecule amenable to easier synthesis.

Monlunabant is yet another drug ready to fight the battle of the bulge. Anyone who has ever tried cannabis will agree that it boosts the appetite. This is because it activates cannabinoid receptors located throughout the body. Novo Nordisk's monlunabant is an oral cannabinoid receptor inverse agonist, which means that it binds to a receptor and produces an effect opposite to an agonist. Since cannabidiol in cannabis increases appetite by fitting into a cannabinoid receptor, monlunabant has the opposite effect. It remains to be seen if these new drugs pan out, but the potential is there to have an impact on the obesity crisis.

There is no question that many type-2 diabetics have benefited from the use of GLP-1 agonists and these drugs have also resulted in significant weight loss in people who have long struggled with obesity. But these are serious medications with a variety of possible side effects, something that may be obscured by the vigorous advertising on U.S. television that shows Ozempic users happily doing yoga or playing pickleball. In Canada, such advertising is not allowed, but ads that advise viewers to ask their doctor about Ozempic are. And when you ask, make sure to ask about the risks and benefits. If you are looking for a miracle, you will have to look elsewhere.

"NATURE'S OZEMPIC." REALLY?

The message was spreading like wildfire on TikTok. Berberine, a compound found in plants such as barberry, golden seal, and meadow rue, is "nature's Ozempic." Why the comparison with Ozempic? Because, as we have seen, semaglutide, the active ingredient in Novo

Nordisk's anti-diabetes drug, turned out to have an unanticipated side effect. Weight loss! That unforeseen outcome stimulated randomized, double-blind clinical trials in which semaglutide helped participants lose as much as fifteen kilograms over sixteen months. Given the number of people who dream of reducing their girth, and the rather dismal historical performance of weight control drugs, it isn't surprising that Ozempic blasted its way into headlines. Neither is it surprising that various Ozempic wannabees have tried to jump on the drug's coattails. Since semaglutide is expensive, running at around $1000 for a month's supply, there's a sizable population ready to be snared with an offer of a cheaper version. Especially if it can be connected to historical use in Ayurvedic and Traditional Chinese Medicine and anointed with the magical marketing term of "natural."

Berberine may indeed be "natural," not that this has any relevance. Ingest some "natural" aflatoxin B, a compound produced by a mold that can be found on some improperly stored grains, and you may quickly lose weight. But it will be because of the cancer sparked by this "natural" toxin. Neither is berberine any version of Ozempic, which is an analogue of glucagon-like-peptide-1 (GLP-1), a natural hormone that regulates blood sugar and helps people feel full. Berberine has nothing to do with GLP-1. How then has this compound been dredged out to become the darling of the TikTok crowd?

It is hard to identify ground zero for berberine as a supposed weight control substance, but a reasonable guess is that someone seeing the Ozempic bandwagon accelerate as it rolled by had the idea of scavenging through the scientific literature to find some substance that could be passed off as a bargain-basement relative. That bit of fool's gold was likely found in a review paper published in *Clinical Nutrition ESPEN*, an Iranian journal with an impact factor of 0.67, which means that papers published in the journal are hardly ever cited by other scientists.

In any case, the article in question was a review paper by Iranian scientists of all the publications they could find in the scientific literature that had in any way related the use of berberine to control weight.

They found twelve, all in low impact journals. Nine were in Chinese publications, and one each in an Italian, Mexican, and Iranian journals. The studies spanned anywhere from one to four months and recorded an average weight loss of about 4.5 kg. Now for the buts.

Significantly, none of the studies tested berberine in subjects whose only problem was being overweight. The studies involved subjects, rather small numbers in each case, who had diabetes, non-alcoholic fatty liver disease (NAFLD), polycystic ovary syndrome (PCOS), coronary disease, or had had kidney transplants. The dosage of berberine ranged from 300 to 1,500 mg a day, and in some studies other drugs were also used. Basically, it is not possible to tell whether any weight loss that occurred was due to berberine or disease. Furthermore, there was no dose-response relationship documented, so in no way can one conclude from these studies that berberine is "nature's Ozempic."

But let's not throw out the baby with the bathwater. Berberine may not help shed pounds, but it may yet make it into pharmacies as an effective medication. At least, some derivatives of berberine may. Drug discovery is a tough and risky business with roughly only one in five thousand compounds tested by researchers ever reaching clinical application. Pharmaceutical companies and academic researchers are constantly searching for candidate molecules that have therapeutic potential, and over the years various compounds found in botanicals used in traditional medical systems have been turned into drugs or have been the base for synthetic variants. Morphine extracted from the opium poppy, colchicine from the autumn crocus, as well as the first semi-synthetic drug, aspirin, made from naturally occurring salicin isolated from the bark of the white willow, are classic examples. Today, scrutinizing traditional herbal remedies for clues that may aid in drug development is a fertile area of research.

Plants that contain berberine, although usually in combination with other botanicals, have been used in traditional Asian medicine for thousands of years for intestinal problems, diarrhea, parasite infection, lowering of body temperature, toothache, and malaria. Researchers' fancy is always tickled when some substance has a history of such a long use and for such diverse conditions. Indeed, in the last few decades numerous papers dealing with berberine have been published. Treating cell cultures, animals, and, in a few cases, humans, with berberine extracted, using alcohol, from plants has produced some interesting results that include antioxidant, anticancer, cholesterol lowering, antifungal, antibacterial, anti-inflammatory, and memory-enhancing effects. Intriguing, but of course there are numerous compounds that have shown such potential in preliminary studies that have ended up on the large refuse heap of the pharmaceutical industry.

Perhaps the most interesting laboratory finding, given the rising global tide of type 2 diabetes, is the control that berberine may exert over blood sugar. But there is a problem. Berberine is virtually insoluble in water and has low intestinal absorption, which means it has poor bioavailability. That is why current research focuses on improved delivery systems. These include nanoparticle formulations that encapsulate berberine to improve bioavailability and synthetic modifications of the molecule to increase solubility. Because of berberine's poor bioavailability, supplements on the market are likely to be useless. However, some derivative of berberine may yet make it to the physician's prescription pad. But it won't be for weight loss. And don't hold your breath.

GARLIC, VAMPIRES, AND BLOOD

The night is cold and misty. The eerie silence is broken only by the occasional howl of a wolf. A perfect night for vampires! Prepare the crosses, sharpen the wooden stakes, and, most important, start digging up the garlic! Transylvanian peasants will tell you that vampires just hate the stuff! Where do they get these ideas? Well, there just may be some science behind this engaging bit of folklore.

The basis of the vampire legend may be a rare inherited disease known as iron deficiency porphyria. In this condition the body cannot properly utilize iron to form the essential oxygen-carrying compound, hemoglobin. This leads to a waxy pallor, extreme sensitivity to sunlight, and receding gums, all characteristics of a vampire. But what's the garlic connection? It turns out that allyl disulfide, one of many compounds found in garlic, activates an enzyme which destroys old blood cells by removing iron from hemoglobin. So garlic can destroy whatever viable hemoglobin the vampire still possesses. The trouble is you have to get the garlic into the vampire's bloodstream. Inviting the vampire to sit down for a bowl of garlic soup will not work. Their diet is limited to human blood. Perhaps a prospective victim, preferably a young lady in a nightgown, can be convinced for the greater good to load up on garlic and submit to the vampire's passion. When he drinks her blood, he'll get a good dose of garlic and hopefully succumb to its effects. In this way the messy business with the stake through the heart can be avoided.

While garlic may be poisonous to vampires, it has a decidedly opposite effect on humans. In fact, it is the most commonly used "health food" in the world. Over the years, in addition to its well-known anti-vampiric effect, garlic has been recommended as a treatment for a variety of ailments as well as a general body tonic. The Egyptians fed large doses to their slaves to keep them strong and healthy, the ancient Greeks claimed that garlic would "open obstructions" in the body, and in India lotions made from the bulb have long been used for the washing of wounds and ulcers. More recently, an American food faddist with the intriguing name of Adolphus Hohensee urged his followers to "cure low blood pressure, inhibit germs, and cleanse the blood and intestines" by using a clove of garlic as a suppository at night. The taste of garlic in the mouth upon waking up was "proof" that the miraculous substance had worked its way through the body and had indeed cleansed the system.

What does modern science say? Without a doubt, garlic juice does have an antibacterial effect. The juice was used in both World Wars as an antiseptic for the prevention of gangrene. The active ingredient appears to be allicin, the compound responsible for the odor of garlic. Allicin is the garlic's natural protective factor, guarding the bulb against attack by fungi, animals, and insects. Indeed it is well known that mosquitos shy away from garlic eaters. But then again, so do people. Allicin has been clinically tested as an antibiotic, but it has never made it to the marketplace due to the substance's antisocial odor and the availability of far better antibiotics. It is noteworthy, though, that four condemned criminals who were forced to bury the dead during an outbreak of the plague in France in the eighteenth century proved to be immune to the disease. Credit was given to the concoction of macerated garlic in wine which they consumed. Vinaigre des quatre voleurs is still available in France today.

The garlic bulb is virtually odor free until it is cut or crushed. It is this physical manipulation that destroys cells and leads to the liberation of the enzyme alliinase, which converts the odorless compound alliin to the odiferous allicin. The largest amount of allicin is formed when garlic cloves are peeled and crushed. This is why noted cooks do not use a garlic press, preferring instead to slice the cloves to impart only a hint of flavor. On the other hand, if we are looking at the potential medicinal value of garlic, we want the allicin, so crushing the cloves is in order. But don't cook these right away. It takes about fifteen minutes for the liberated enzyme to do its job. If you toss the garlic into the saucepan right away, the heat will destroy the enzyme and, with it, some of the medicinal potential.

The study of garlic as a medicine is complicated by the fact that besides allicin, the bulb contains literally hundreds of compounds. Two of these are of particular interest because of their ability to interfere with the clotting action of blood. Allyl methyl trisulfide, a substance notoriously difficult to isolate in quantity from garlic bulbs, can now be synthesized in the laboratory. A second compound, a non-smelly component of garlic known as ajoene, appears to have an even stronger antithrombotic effect. It also lends itself to laboratory synthesis and has potential to find a place alongside aspirin, heparin, and coumadin as a commonly used anticoagulant medication. The presence of these compounds in garlic extracts, powders, pills, and lotions marketed in health food stores is questionable. If there is to be a protective effect, it is to be found in the consumption of fresh garlic, and lots of it. But then there is the issue of a garlicky smell, feeling hot, burning urine, heartburn, flatulence, and belching. On the other hand, garlic lovers claim increased energy and enhanced sexual desire.

So blood and garlic are obviously related, which is where we started our story. As far as scientific evidence for protection against vampires goes, I cook with a lot of garlic and have never felt a vampire breathing down my neck at night.

THE STORY OF ACETAMINOPHEN

Loudspeakers atop patrol cars in Chicago boomed out a terrifying warning: "Do not consume any Extra Strength Tylenol capsules you may have purchased!" A media blitz about the potential lethal effects of taking the painkiller followed. The year was 1982, and seven people had just died of cyanide poisoning!

It didn't take authorities long to find the common link. All the victims had taken capsules of Extra Strength Tylenol! The manufacturer, Johnson & Johnson, quickly recalled thirty-one million bottles, but it soon became clear that the company was blameless. The tainted capsules were limited to a few bottles that had been purchased in the Chicago area. Apparently, some wretched soul had managed to open capsules and insert potassium cyanide before placing the bottles with their poisoned contents on pharmacy shelves to be purchased by unfortunate customers.

Despite a massive police investigation, the murderer was never found, and no motive was identified. Early suspicion fell on James Lewis who had sent a letter to Johnson & Johnson in which he claimed responsibility and demanded a million dollars to stop the poisonings. But it was determined that he was just trying to capitalize on the sordid affair and had nothing to do with the cyanide-laden capsules. Lewis ended up being sentenced to ten years in prison for extortion.

While this tragic episode had nothing to do with Tylenol's active ingredient, that ingredient merits some scrutiny, especially given that it is the most widely used medicine in the world. Tylenol's name derives from aceTYLaminophENOL, also known as acetaminophen in North America and paracetamol elsewhere.

Back in 1884, two young German physicians, Arnold Cahn and Paul Hepp, made a serendipitous discovery when treating a patient who was infested with worms and was running a fever. On the advice of their mentor, Professor Adolf Kussmaul, the two doctors tried administering naphthalene, believed at the time to be effective against worms. It did nothing for the worms, but it did reduce the patient's fever, an effect that had not previously been described.

Intrigued, Cahn and Hepp wondered whether the naphthalene they purchased may have contained some other substance as well. Tracking down the source, they discovered that the pharmacy had made a mistake and had provided acetanilide instead of naphthalene. In those days, pharmacies not only dispensed drugs but also stocked chemicals used in research laboratories. One of these was acetanilide, a derivative of aniline, a compound that had recently been isolated from coal tar and used in the production of the novel synthetic dyes that were all the rage. It now occurred to Cahn and Hepp that acetanilide might be a useful fever reducer. As luck would have it, Hepp's brother worked as a chemist at the Kalle pharmaceutical firm which was willing to investigate the compound. Finding that it did indeed reduce fever, acetanilide was put on the market as "Antifebrin."

Unfortunately, it wasn't long before a problem cropped up. Users complained of "blueish skin," a sign soon identified as methemoglobinemia, a condition in which oxygen transport by hemoglobin is impaired. This precipitated a search for safer alternatives. As is still common today, such a search involves making alterations in the molecular structure of a compound, hoping to eliminate side effects while maintaining its therapeutic potential. In 1887, the German company Bayer modified acetanilide to produce phenacetin which did not cause methemoglobinemia. The same year, physician Joseph von Mering tried another derivative of acetanilide that had been synthesized by Harmon Morse at Johns Hopkins University. This was acetaminophen, which worked well as a pain reliever and fever reducer, but von Mering claimed that, unlike phenacetin, it had a potential to produce methemoglobinemia. For this reason, acetaminophen languished on laboratory shelves until 1947 when researchers discovered that both acetanilide and phenacetin were converted into acetaminophen in the body and the latter was the actual active agent! Furthermore, von Mering had been wrong, even large doses of acetaminophen did not cause methemoglobinemia in rats.

In the meantime, phenacetin had been linked with kidney problems, and its competitor, aspirin, with stomach irritation and Reye's syndrome, a rare children's disease. This set the stage for the promotion of acetaminophen as a safer drug and led to its mass marketing as Tylenol in 1955.

Acetaminophen turned out to be a popular pain reliever and fever reducer, safe when taken in the proper dose, which for an adult is a maximum of six 500 milligram pills a day. Exceeding this dose can lead to serious problems due to a metabolite of the drug being toxic to the liver. (N-acetyl-p-benzoquinone imine, if you must know.) Just twenty pills of acetaminophen taken over a short period can be lethal!

Normally, the toxic metabolite is mopped up by glutathione, one of the body's prime detoxicating agents, but it cannot handle an overdose. Indeed, acetaminophen overdose is the predominant cause of acute liver failure in industrialized countries, and is the prime reason for liver transplants. Luckily, the vast majority of the 60,000 emergency room visits a year in North America due to acetaminophen toxicity do not have such dire consequences thanks to emergency rooms stocking an antidote, N-acetylcysteine (NAC), which can help replenish glutathione.

Today, the global annual acetaminophen market is an astounding nine and a half billion dollars, fueled by heavy promotion that emphasizes the lack of gastrointestinal side effects associated with aspirin and other non-steroidal anti-inflammatory drugs (NSAIDs). While the concern about liver toxicity has to be recognized, there is no worry about cyanide-tainted Tylenol. A consequence of the Chicago tragedy was the development of tamper-proof packaging that features a plastic seal, an aluminum foil cover, and a pop-up cap. But we now face the challenge of opening the bottle, a struggle that can exacerbate the headache we are looking to treat.

HANS SELYE AND STRESS

In 2000, Canada issued a set of four stamps honoring "medical innovators." These included Sir Frederick Banting, who pioneered the use of insulin in diabetes; Dr. Maud Abbott, expert in congenital heart disease and one of the first women to graduate from medical school in Canada; Dr. Armand Frappier, Quebecois promoter of vaccination; and Dr. Hans Selye, the "father of stress research." The Selye stamp featured his portrait; a partial molecular structure of a steroid molecule; the letters *A*, *B*, and *C*; and the word *STRESS*.

I purchased the set, mostly because I have a special regard for Dr. Selye. Back in the early 1960s, my parents dragged me to a public lecture he gave at McGill University, not because they were particularly interested in the topic but because Selye, like us, had a Hungarian background. I don't remember much of what he said. Talk of stress hormones was surely above my head at the time. But he did make a point that became etched in my mind. "You could leave here today," he began, "get back a tad late to your car, and find a parking ticket on the windshield. Then you have a choice. You can rant and rave about the unfairness of it all, or you can just accept that you should have looked at the time more carefully. In either case," he concluded, "the cost to your wallet will be the same, but not the cost to your health." The message was that anger could trigger biochemical changes with adverse effects on health.

I have often recalled that example as the spark that ignited my interest in the connection between the body and the mind and my fascination with the career of Dr. Selye. He began with his earning diplomas both in medicine and organic chemistry in Prague, and it was during his medical school days that he made an interesting observation. Besides exhibiting symptoms of their particular disease, patients also shared identical symptoms regardless of their specific condition. They all had coated tongues, diffuse joint aches and pains, intestinal disturbances, as well as loss of appetite and muscular strength. Selye wondered if there was a scientific explanation for the "syndrome of just being sick."

Years later, as a young researcher in biochemistry at McGill, he would propose an explanation based on his studies with rats. Selye described this work in a landmark paper, "A Syndrome Produced by Various Nocuous Agents," published in 1936 in the prestigious journal *Nature*. That paper is widely regarded as the seminal work that would eventually lead to Hans Selye's name being forever associated with stress.

At McGill his research focused on hormones and involved injecting extracts of cattle ovaries into rats. He reported that the thymus gland, spleen, and liver shrank, while the cortex of the adrenal gland enlarged. Acute erosions appeared in the animals' digestive tract, their body temperature fell, and there was a loss of muscular tone. No great surprise here, but Selye was astonished to find that injections of sublethal doses of other "nocuous" agents such as atropine or formaldehyde produced the same changes! Ditto for subjecting the rats to excessive muscular exercise on the treadmill, or exposing them to cold, heat, or surgical injury. It didn't matter to what trauma the rats were subjected, the response was the same. Selye realized that he had chanced upon an experimental replication of the "syndrome of just being sick" he had observed in humans back in medical school.

As he pursued this research, Selye discovered that continued treatment with relatively small doses of toxins or other stressors results in the animals building resistance to the insult. Beginning around forty-eight hours after the injury, the appearance and function of their organs returned to normal. But if the low-level insults were continued for months, the animals lost their ability to adapt, and the initial symptoms returned, often with a vengeance. Selye called this the "exhaustion stage."

He went on to describe that stress syndrome was characterized by three stages, explaining why the letters *A*, *B*, and *C* appear on the stamp. The initial, or alarm stage, is the fight-or-flight response, first described in 1915 by Harvard physiologist Walter Cannon. The second stage is adaptation to the stressor, and in the third stage, as a result of continued stress, adaptation energy runs out and health suffers.

Eventually, the biochemistry involved was worked out. In the first stage, adrenaline and cortisol are released from the adrenal glands causing a quickening of the pulse and respiration so that more oxygen is delivered to tissues. Blood sugar rises to energize muscles, and blood platelets aggregate as the body anticipates possible bleeding from an injury. At first, cortisol sustains the stress reaction, but then slows it down so the body returns to normal. However, if cortisol stays elevated, it suppresses the immune system, keeps blood sugar and blood pressure elevated, and causes what Selye called "wear and tear" on the body.

Selye proposed that people can control how they adapt to stress and thereby can exert control over cortisol levels. He gave the example of a drunk, clearly unable to do any harm, hurling insults at someone. That someone has a choice. Walk away, ignoring the insults, or get all riled up and start some sort of confrontation. In the latter case, adrenaline and cortisol will kick in, and if there is some faulty biochemistry, the second stage of the reaction, the adaptation stage, will be skipped and the exhaustion stage with its dire consequences will quickly set in. The result may be a cortisol-triggered heart attack that Selye claimed was caused by choosing the wrong reaction. He summed it all up by saying that stress is largely in the eye of the beholder; it is not so much what happens to you as it is how you respond. You have to decide whether fight or flight or just relax is appropriate. In yet other words, don't sweat the small stuff. As Kenny Rogers reminded us: "You've got to know when to hold 'em, know when to fold 'em, know when to walk away, and know when to run . . ."

HAHNEMANN, HOMEOPATHY, AND CHEMISTRY

I use a lot of visuals in my presentations since as they say, a picture is worth a thousand words. These days the go-to source for such images is the internet, so that is exactly where I went when I was looking for a picture of the grave of Samuel Hahnemann, the founder of homeopathy. Curiously, one of the pictures that came up was of a coffee mug with a portrait of Hahnemann. Why was that curious? Because in 1803 Hahnemann published his "Treatise on the Effect of Coffee" in which he claimed that coffee was the cause of impotence, sterility, rickets, stammering, melancholy, malicious envy, and insomnia. He was at least right about insomnia.

Seven years before his attack on coffee, Hahnemann had introduced the famous dictum "like cures like" in his "Essay on a New Principle for Ascertaining the Curative Power of Drugs." At the time, powdered cinchona bark was used as a treatment for malaria, but there was always a question of how much to use. Hahnemann, acting as his own guinea pig, took increasing doses to see how much a patient could safely take. As he increased the dose, he began to experience the symptoms of malaria. At that moment the concept of homeopathy was born! If a patient had an illness, the cure was a substance that, if given to a healthy person, produced symptoms similar to that experienced by the sick person. To many, that seemed reasonable given that Hahnemann cited Edward Jenner's use of vaccination with cowpox to prevent smallpox. Indeed, injection of a small dose of cowpox produced symptoms similar to smallpox and offered protection against that dreaded disease.

However, controversy was stirred when Hahnemann suggested that doses so small that they would produce no symptoms would still be effective in treating disease. His dilutions were so extreme that it seemed inconceivable that they could have any therapeutic effect, but Hahnemann had a rationale. He claimed that the diluted preparations retained their therapeutic power if after each dilution the solution was violently shaken. This "potentization" resulted in a "dematerialized spiritual force" that produced a cure. Even at the time, when science was still on a shaky footing, this was deemed to be nonsense by physicians. Hahnemann was further mocked when he went on to claim that nearly all chronic diseases were caused by scabies.

Today, the scientific view of homeopathy is that nonexistent molecules cannot cure existing disease, but homeopaths can offer some comfort to patients by triggering a placebo response. Also, the conditions for which patients consult homeopaths are generally ones that are transient and disappear spontaneously or are cyclical. If taking a homeopathic "remedy" is followed by remission, the remedy gets the credit.

Given that I have forged a career on separating sense from nonsense, you might think that I would be keen to dismiss Samuel Hahnemann as a misguided simpleton. Not so. We have to remember that at the time orthodox medicine did not have much to offer beyond bloodletting, purging, and potentially toxic substances such as mercury, strychnine, or opium. Hahnemann had concluded that such practices just tortured patients, and he looked for kinder, gentler treatments. While he was wrong about his homeopathic treatment having a physiological effect, he did offer patients an alternative to the brutal medical practices of the times. Little wonder that patients were happier to take a sugar pill infused with a solution that contained nothing than to be purged or be drained of their blood.

There is yet another reason that I have a special regard for Hahnemann, and it has nothing to do with his medical practice. In 1788, before he hatched his like-cures-like scheme, he wrote an article that began thus: "I do not know whether I am mistaken, but it seems that one can obtain more truths, important to Humanity, from Chemistry than from any other Science." At the time, Hahnemann was a struggling physician and was more active as a chemist than as a doctor! He had become enamored of chemistry while studying medicine at the University of Leipzig where he had taken a course in experimental chemistry from noted physician and chemist Johann Gottfried Leonhardi.

In the article that began with his exaltation of chemistry, Hahnemann described his experiments to develop a test for lead and iron in wine. Lead was sometimes added to wine to act as a preservative as well as to increase sweetness, while iron ended up in wine from nails in casks. Lead was known to be poisonous, but the problem was that the current test for lead, based on adding a solution of arsenious sulfide to produce a dark precipitate of lead sulfide, also gave a positive result for innocuous iron. Hahnemann discovered that an acidified solution of hydrogen sulfide added to wine yielded a precipitate with lead but not with iron!

To acquire the chemicals he needed, Hahnemann visited the local pharmacy. There was an added inducement for frequent visits since he fell in love with the pharmacist's stepdaughter whom he went on to marry. They were married for forty-eight years and had eleven children. Five years after her death, at the age of eighty, Hahnemann took the plunge again with Marie Melanie d'Hervilly, who was forty years his junior and had come to consult him about her health. She convinced him to move to Paris where he developed a large following and was consulted by patients from around the world. Melanie followed in her husband's footsteps and became the first female homeopath but was not welcomed by the homeopathic community. There were accusations of delaying publication of Hahnemann's final work as she held out for a more lucrative offer.

Samuel adored Melanie. Speaking of whom he would trust to carry on his work, he proclaimed, "I have long sought a man and have found him in my wife." His expressed desire was that they be buried alongside each other, and indeed they do lie together at Pere Lachaise Cemetery in Paris. In 1900, the International Homeopathic Congress erected a beautiful granite monument on the gravesite, but nowhere is there a mention of Melanie Hahnemann, Samuel's beloved wife who helped propel him to fame and fortune. Neither is there any mention of his contributions to chemistry.

Now you see why I was looking for the picture of Hahnemann's grave. As far as the coffee mug with a portrait of Hahnemann goes, needless to say, I ordered it.

THE UPS AND DOWNS OF OXYCONTIN

In the mid-nineteenth century, the British East India Company had a lucrative business smuggling opium from India into China. The Chinese government had been aware of the addictive potential of smoking opium and had made the importing of the substance illegal. China's attempt to block its ports from ships carrying opium led to the First Opium War in 1834 with the British navy breaking the blockade. Nine years later, China's seizing a British ship triggered the Second Opium War which resulted in a quick British victory and the legalization of the opium trade.

Today, we are facing another type of opium war. The opioid epidemic is destroying lives with fingers being pointed at one specific drug, OxyContin, as being responsible for causing much of the misery. The story is a complex one, but there is one inevitable conclusion. Purdue Pharma, the company that introduced OxyContin to the market in 1996, has a closet full of skeletons. Deceptive marketing, the spread of misinformation, and concealment of the addictive nature of the drug were all part of a scheme to boost sales. And that scheme netted Purdue Pharma billions of dollars.

Opium is the white resin that seeps out of the bulb of the opium poppy when it is cut with a knife. It has a complex chemical composition but the physiological effects that can include euphoria, pain relief, and sleep are mostly attributed to morphine, codeine, and thebaine. The name *morphine* derives from Morpheus, the Greek god of sleep, and unfortunately, in the wrong dose, that sleep can be permanent.

The opium poppy was cultivated as early as 3400 B.C. by the Sumerians in Mesopotamia who referred to it as the "joy plant." The Assyrians learned about the poppy from the Sumerians and passed the knowledge on to the Egyptians. By the time King Tutankhamun ascended to the throne around 1300 B.C., Egypt had a flourishing trade in opium. Around 460 B.C. Hippocrates acknowledged the sleep-inducing properties of opium, but curiously there is little mention of the substance until the sixteenth century when Paracelsus introduced "laudanum," a mixture of opium, citrus juice, and powdered gold as a painkiller. By the nineteenth century, smoking opium and taking morphine, isolated from opium in 1803 by German chemist Friedrich Sertuerner, had become popular medically and recreationally. It had also become clear that opium was addictive.

The challenge was to find an alternative to morphine that retains its pain-killing properties but eliminates its addictive potential. One approach was to make small changes in the compound's molecular structure and hope for the desired outcome. In 1895, Heinrich Dreser at the Bayer company reacted morphine with acetic anhydride and came up with diacetylmorphine which the company named heroin. It was deemed to perform "heroically" in eliminating pain without causing addiction. Wrong. Heroin turned out to be even more addictive than morphine.

In 1916, Martin Freund and Edmund Speyer at the University of Frankfurt tried a different approach. Instead of morphine, they chose thebaine, another constituent of opium, as their starting material. Reacting this compound with hydrogen peroxide, then with hydrogen gas, produced oxycodone. The new drug didn't solve the addiction problem, but it did produce pain relief at a dose equivalent to morphine but was less sedating. During the Second World War, oxycodone became Germany's main pain reliever on the battlefield and Theodor Morell, Hitler's physician, documented that he gave the Fuhrer repeated injections of Eukodal, the trade name of oxycodone, after he was injured in an assassination attempt.

By 1937, oxycodone had been introduced in the U.S. but did not make much of a mark until the 1980s when a little-known pharmaceutical company that had been established in 1952 by three physician brothers, Arthur, Mortimer, and Raymond Sackler, introduced a slow-release form of morphine sulfate under the name MS Contin. The *contin* was a reference to *continuous*. The idea was that the active ingredient would be released slowly, allowing pain relief over a longer period. This was accomplished by coating the pill as well as the particles of the active ingredient it contained with varying thicknesses of proprietary resins and acrylic polymers that allowed for passage through the stomach into the small intestine from where oxycodone would be slowly absorbed. MS Contin sold well, but its patent protection was running out, and Purdue Pharma needed to plug a new drug into the pipeline.

The company decided to apply the same technology that had been used in MS Contin to oxycodone and came up with OxyContin which was introduced in 1995. An aggressive marketing campaign was launched with a focus on OxyContin providing twelve hours of relief instead of the four hours provided by most opiates. Furthermore, because of the slow release, there would be no euphoria and therefore a smaller likelihood of addiction. There was no evidence for either of these claims! In fact, a Purdue-sponsored trial in 1989 had shown that after abdominal surgery more than a third of patients found that OxyContin's relief of pain lasted only eight hours.

Arthur Sackler was a whiz at marketing and had already made a fortune from a medical advertising company he founded. His "trick," of focusing on seducing doctors as well as patients, revolutionized the industry. Armies of salespeople marched into doctors' offices claiming that OxyContin was ideal for any kind of pain, including back aches and wisdom teeth extractions. And they underlined that less than one percent of users became addicted to the drug. This was blatantly untrue. It wasn't long before it became clear that many patients were showing withdrawal symptoms, often just six hours after taking OxyContin. This made them take a second dose, a recipe for addiction. Deaths from respiratory arrest among OxyContin users were becoming common, but Purdue Pharma insisted that it was due to people abusing the drug by grinding up and snorting the pills. This was not true. Addiction and deaths were also seen in patients who were using the pills as directed by physicians.

Purdue Pharma's replacement of OxyContin by OxyNEO, which was supposed to thwart abusers by making the pill difficult to crush and snort, was in fact just a way to get patent protection for a new drug as the patent protection for OxyContin was coming to an end.

The boom began to descend on Purdue Pharma in the early 2000s with numerous lawsuits by patients and state governments being launched. Finally, in 2019, after paying billions in fines and settlements, the company filed for bankruptcy, but the settlement was overturned by the U.S. Supreme Court. In March of 2025, Purdue Pharma L.P. filed a new Chapter 11 Plan of Reorganization, pledging to create a new public benefit company, 100 percent devoted to improving the lives of Americans. Purdue officials have always maintained that the problem was not the drug, but people who abused it. The problem was people all right, people who worked for Purdue Pharma and spread misinformation.

HARRY POTTER AND THE MANDRAKE

"The mandrake forms an essential part of most antidotes, it is also, however, dangerous. Who can tell me why?" That was the question Professor Pomona Sprout posed to her Herbology class at Hogwarts. Harry Potter had no idea, but Hermione Granger did. "The cry of the mandrake is fatal to anyone who hears it." Professor Sprout acknowledged that the answer was indeed correct and proceeded to hand out earmuffs to the class as she prepared to demonstrate the repotting of mandrake plants. They were to be used to make a potion "to return people who have been transfigured or cursed to their original state."

J.K. Rowling did not pull the mandrake story out of thin air. She had quite an extensive knowledge of herbs, mostly gained from Nicholas

Culpeper's seventeenth century classic *The Complete Herbal*, which also stimulated her to explore the rich folklore of plants. And when it comes to the root of the mandrake plant, a member of the nightshade family, the folklore is indeed rich. But it is also speckled with some real science.

One of the earliest mentions of the effects of mandrake takes us back to the book of Genesis in the Old Testament. Rachel, who is barren, agrees to let her sister Leah "lie" with her husband Jacob in exchange for some mandrake that Leah's son had found. The plant's root was supposed to impart fertility, and as the story goes Rachel soon gave birth to Joseph and Benjamin. The fertility saga may be rooted in the shape of the underground stem of the plant which, with a bit of imagination tossed in, resembles the human body. Much later, in the Middle Ages, the myth would be resuscitated with the popularization of the Doctrine of Signatures. The suggestion was that God had placed a "signature" on plants by having them resemble parts of the body to give a clue about their potential medicinal use. John Donne, the British poet, apparently bought into fertility myth, "Go and catch a falling star, Get with child a mandrake root," he penned in his famous poem "Song."

While the Doctrine of Signatures has no scientific merit, plants of course can harbor a multitude of physiologically active compounds. The mandrake root contains atropine, scopolamine, and hyoscyamine along with a number of other alkaloids that can in a sufficiently high dose cause effects that range from drowsiness and hallucinations to respiratory failure and death. As early as the first century AD, Dioscorides in his famed *De materia medica* suggested that a decoction of mandrake in wine would take away the pain of snake bite and also "make patients insensitive to incisions and cauterizations."

Curiously, there seems to be no further mention of the effects of mandrake until around the tenth century when descriptions of a spongia somnifera, or soporific sponge, began to appear. Although there were several versions, the basic method was to soak a sea sponge in an extract of herbs that always included mandrake as well as henbane and the opium poppy. The sponge would be allowed to dry, and when needed to induce sleep, it would be moistened and placed under the nostrils of the patient.

The herbal extracts could possibly have induced sleep if ingested, but sniffing a soporific sponge would not have worked. Nevertheless, the association of mandrake with sleep caught the imagination of Shakespeare. "Give me to drink mandragora, That I might sleep out this great gap of time," exclaims Cleopatra in *Anthony and Cleopatra*. Mandrake was also an ingredient in "witch's ointment," first mentioned in the fifteenth century. This was supposedly smeared on the skin to give the sensation of flying through its hallucinogenic effect.

Due to its properties, both mythical and real, mandrake was in great demand in medieval times. Perhaps the most famous fable about the plant was spawned by growers' attempts to thwart theft. The root was said to alter a terrifying scream when pulled from the ground, a scream that was potentially fatal to humans. The practice therefore was to tie a hungry dog to the stalk and entice it with food placed some distance away. When the dog went for the food, it uprooted the mandrake while its master watched safely from afar. Unfortunately, the dog did not survive the mandrake's scream.

Shakespeare knew about this myth, as evidenced in *Romeo and Juliet*: "And shrieks like mandrakes torn out of the earth, That living mortals, hearing them, run mad." The legend also inspired J.K. Rowling's weaving it into several scenes including one in *Harry Potter and the Chamber of Secrets*. Hogwarts students were being petrified by Basilisk, a giant reptile, and mandrake was used to make a restorative draught to bring the prospective wizards and witches out of their solidified state.

Today, visitors to the Warner Brothers Studios in London, where the Harry Potter movies were filmed, can engage in an interactive display that allows them to uproot a mandrake and hear it scream. No fatalities have been recorded.

Modern science has stripped mandrake of its enchanting mythology and has found no safe medical use for the plant, although some herbalists still recommend its use without any supportive evidence. There is, however, one remnant of mandrake's magical history.

The first superhero, predating Superman, was Mandrake the Magician, created by cartoonist Lee Falk in 1934. His top hat, cloak, and wand were endowed with magical powers that he used to fight villains. The character may well have been based on Leon Giglio, a real-life Canadian magician whose mustache and attire resembled the cartoon version. Giglio went on to capitalize on this connection by legally changing his name to Mandrake. For over half a century he entertained the public in stadiums, theaters, night clubs, and on television with his illusions, escapes, and mind-reading act. This Mandrake, unlike the root, was really magical.

DIETHYLENE GLYCOL IS NOT FOR CONSUMPTION

Every year, Germany's Gesellschaft für deutsche Sprache, a government sponsored language society, chooses a word that dominated the news that year. In 2020, the choice was easy. *Coronavirus*. This was followed in 2021 by *wellenbrecher*, a word describing measures which stop new waves of COVID-19 infections.

Back in 1985, the society chose the word *glycol*. Technically it should have been *diethylene glycol*, but glycol was easier to splash around in headlines. And there were headlines galore! All about Germans being poisoned by adulterated Austrian wines! The Austrian wine scandal rocked Germany, put a huge crimp into the sale of Austrian wines, and resuscitated discussions about diethylene glycol, a chemical that had shaken America in 1937 with its lethal effects.

French chemist Adolphe Wurtz first synthesized diethylene glycol in 1860 and found it to be a useful solvent for chemical reactions. Nothing significant about the substance appeared in the scientific literature until 1937 when it catastrophically reared its head. Sulfanilamide pills had been introduced in Europe as an effective treatment for some infections, but Americans preferred dosing themselves with liquids. Chemist

Harold Watkins at the S.E. Massengill Company of Tennessee was given the task of finding a solvent for the drug and, after scrutinizing his shelves, came up with ethylene glycol. It readily dissolved sulfanilamide, and when Watkins tasted a drop, he found it sweet. And that was it. There was no requirement at the time for safety testing, and off to the marketplace it went. Within a few months, 107 people had died from diethylene glycol poisoning! The tragedy led to President Roosevelt signing the 1938 Food and Drugs Act that required drugs to be proven safe before being sold.

Although the sulfanilamide affair received widespread publicity in America, it seems accounts about the toxic effects of diethylene glycol did not make it to Austria. And therein lies a captivating story. Austria had become famous for producing high-quality, full-bodied sweet wines from grapes left on the vines for a long time. These wines were particularly favored in Germany, and a number of Austrian producers signed contracts with West German importers to provide the highly prized varieties. However, in the early 1980s, climate intervened and the grapes could not achieve the desired ripeness. The wine made from them was thinner and less sweet. This would not pass muster in Germany, and vintners were in a quandary about what to do. One sought help from Otto Nadrasky, a chemist and wine consultant, who suggested doctoring the wine with diethylene glycol, reminiscent of Harold Watkins. Vintners found the results satisfactory, and word spread. The adulterated wines were soon on their way to Germany.

Germans are particular about the wines they drink, and the country already had a system of random testing of bottles for quality. Alarm was raised when a bottle of Austrian wine in a supermarket tested positive for diethylene glycol. It soon became apparent that the problem was widespread, triggering fear and outrage. The Austrian government began a full-scale investigation that prompted producers to try to dump their cache of adulterated wine before they were caught. One dumped 4,000 gallons into the sewer, and the diethylene glycol killed the microbes in the town's sewage treatment plant. Untreated sewage emptied into local streams, killing fish. The man was arrested along with twenty-eight others, including Nadrasky. Whether the latter was aware of the toxicity of diethylene glycol isn't clear, but as a chemist, he should have been. Although many people were made ill by the tainted wine, luckily, unlike the sulfanilamide elixir, the bottles did not contain a lethal dose.

There was still the question of what to do with millions of liters of poisoned wine. A cement plant came to the rescue and modified its ovens to use wine as a coolant. An electrical power plant managed to use the wine as fuel, and the wine mixed with salt proved to be effective for melting ice on roads. The Austrian government passed a law requiring every bottle of wine to be tested for quality and to be sealed with a stamp attesting to having been tested. It took about fifteen years for the Austrian wine industry to recover, but it is now back to producing some excellent wines.

The next twist in the diethylene glycol story takes us to China and the sordid tale of counterfeiters who substituted the sweet-tasting solvent for the safe, but more expensive, glycerin that is used in medications such as cough remedies. In 2005, Wang Guoping, a Chinese tailor looking to make some extra money, purchased some diethylene glycol, relabeled it, and sold it to a pharmaceutical company. The result was a replication of the sulfanilamide disaster. Mr. Wang was caught and sentenced to life in prison. But he is just one of many counterfeiters.

In 2006, with the rainy season approaching, the Panamanian government, anticipating colds and coughs, purchased glycerin from China to produce some 260,000 bottles of antihistamines and cough syrups. The "glycerin" was actually diethylene glycol and ended up killing over a hundred people as proven by the presence of the chemical in exhumed bodies. The shipment was eventually traced to the "Taixing Glycerin Factory," which had never made any glycerin. The "Factory" had bought the diethylene glycol from the same manufacturer as Mr. Wang, the former tailor.

You know that a chemical is notable when it makes it onto *The Simpsons*. A cleverly titled 1990 episode, "The Crepes of Wrath," has Bart being sent to France as an exchange student where his hosts turn out to be unscrupulous winemakers who adulterate wine with "antifreeze." The writers didn't get it quite right. Diethylene glycol is not marketed as antifreeze, but it is found in small quantities in common antifreeze, which is ethylene glycol. That too is highly toxic. A woman murdered her husband in 1995, then her boyfriend in 2001, by putting it in their food. Once more "glycol" made it into the news.

DINITROPHENOL AND THE AMERICAN CHAMBER OF HORRORS

One reporter dubbed it the "American Chamber of Horrors." While there was no actual chamber, there were indeed horrors for the public to see. All they had to do was peruse a traveling exhibit organized by the U.S. Food and Drug Administration in 1933. The display featured products that had the potential to harm consumers yet were out of reach of the Pure Food and Drug Act of 1906. That Act had made misleading claims on labels illegal but did not address claims made in advertisements and did not cover cosmetics. Products such as Koremlu, that claimed to eliminate unwanted hair; Banbar, which was offered as an alternative to insulin; and Nuxated Iron with its promise "to invigorate, rejuvenate, and enhance athletic performance" were readily available to the public. These contained, respectively, highly toxic thallium acetate, a worthless extract of the weed horsetail, and strychnine from the seeds of the nux vomica tree. Something had to be done.

Under the direction of Ruth deForest Lamb, the FDA's chief education officer, and chief inspector George Larrick, the agency mounted an exhibit of some 100 dangerous products that managed to garner extensive media attention when it was presented at the 1933 Chicago World's Fair. The message was that the 1906 Act had to be updated because it did not give FDA the authority to protect the public from deceptive products that were worthless and often toxic.

One of the most dangerous pharmaceuticals targeted by the FDA exhibit was dinitrophenol, which claimed to accelerate metabolism and produce rapid weight loss. The claim was actually valid, but by the 1930s, scientists had discovered that dinitrophenol causes cataracts, low white blood cell counts, and a potentially lethal elevation in body temperature. Yet the FDA was powerless to act because dinitrophenol was sold as a cosmetic and was therefore outside the scope of the 1906 law.

The story of dinitrophenol as a weight-reducing substance begins with Italian chemist Ascanio Sobrero's 1847 discovery that reacting glycerine with nitric acid produces the high explosive, nitroglycerine. This stimulated researchers to investigate nitrating other compounds, and as a result, trinitrotoluene (TNT) and dinitrophenol were produced and found to be useful by the military as explosives in artillery shells. During the First World War, a number of workers in French munitions factories where such shells were produced began to experience weakness, dizziness, excessive sweating, and weight loss.

It was the weight loss that intrigued Stanford University clinical pharmacologist Maurice Tainter, who wondered about the potential use of dinitrophenol as a weight control drug. Indeed, he found that it increased metabolism and led to weight loss without dieting. His published findings were seized upon by devious marketers who unleashed a cascade of products with names such as Nitroment, Nitraphen, and Redusol with claims of quick, safe weight loss without dieting. There was weight loss, but it certainly wasn't safe. Before long, there were reports of cataracts, rashes, nausea, convulsions, liver failure, and death. These dangers were highlighted in the Chamber of Horror display and were a factor in President Franklin Roosevelt signing the 1938 Federal Food, Drug and Cosmetic Act. Finally, the FDA was given the power to remove dangerous substances from the marketplace with one of these being dinitrophenol.

After the Second World War, the compound briefly made it into newspapers again with stories about Soviet soldiers using it to keep warm. This had actually been reported in Soviet medical journals and came to the attention of Russian-born, American-educated Dr. Nicholas Bachynsky, who was translating these journals for the U.S. government. Just like Tainter had discovered, the Russians found that dinitrophenol increased metabolism and causes the body's temperature to rise.

The mechanism by which dinitrophenol produces its effects began to be unraveled in the late 1940s. The drug interferes with the body's production of adenosine triphosphate (ATP) from carbohydrates and fats. It is the dissociation of ATP into ADP (adenosine diphosphate) that produces the energy needed by all living cells. To counter the lack of ATP, the body revs up its metabolic rate to try to convert more fats into energy. That does lead to weight loss but also to the production of excessive heat and potentially fatal hyperthermia.

Like his unscrupulous predecessors, Bachynsky recognized the profit potential of a weight-reducing drug. Although he was aware the FDA had banned the commercial sale of the drug, he thought he could skirt the law by selling it through his private clinic. He purchased industrial dinitrophenol, produced for use in dyes, wood preservatives, and pesticides and processed it into tablets that he began to dispense in 1981 under the name *Mitcal*. Patients were soon reporting a number of adverse reactions to the FDA, resulting in an injunction barring Bachynsky from selling the drug, which he ignored. Charges were brought, including insurance fraud, and the doctor ended up in prison. Here he met another inmate, Dan Duchaine, with whom he hatched an even more troubling scheme.

Upon his release, Bachynsky promoted dinitrophenol as a cancer treatment that would heat up and destroy cancer cells. He enlisted investors in Helvetia Pharmaceuticals, a company he established to develop "intracellular hyperthermia therapy" for cancer. Investors lost their money and discovered that clinical trials had been falsified and the investments had been used to fund Bachynsky's lifestyle. A criminal investigation followed and in 2008 Bachynsky was convicted of fraud and sentenced to fourteen years in prison.

Bachynsky has not been the sole culprit when it comes to sales of dinitrophenol. It is available through multiple internet sites where promoters protect themselves by labeling their product as "not for human consumption." Nevertheless, fatalities among bodybuilders and dieters looking to quickly shed pounds are being reported, about 1,500 since 2007 in the U.K. alone. If the Chamber of Horrors were updated, dinitrophenol would still occupy a prominent place in the display.

FAKE DRUGS, REAL PROBLEMS

We are drowning in fraud. Simply defined, fraud is intentional deception, usually for monetary gain. We have become used to robocalls telling us that we have been subjected to a tax audit and had better call the given number to clear things up. We have learned to dismiss messages from some unfortunate soul supposedly stuck in a foreign country who is in need of funds to get home. Such scams are annoying, but at least not life threatening. The same cannot be said for counterfeit medications, an immense problem with an estimated global annual worth in excess of $100 billion!

We assume that if we take an aspirin tablet that is indicated to contain 325 milligrams of acetyl salicylic acid on the label, then that is exactly what we are getting. That assumption is almost certainly correct since selling fake aspirin is not a lucrative business. But buying Kamagra online is a different story. This pill supposedly contains sildenafil citrate, the active ingredient in Viagra, but costs a fraction of Pfizer's prescription version. It is produced by a company in India and therein lies a problem. Major pharmaceutical companies such as Pfizer, Merck, and Bayer follow exacting manufacturing conditions that ensure their drugs contain the active ingredient in the right dose with their procedures being monitored by government regulatory agencies. The situation is not the same in India where governmental oversight is much more lax. Drugs that have been poorly manufactured or have deteriorated due to improper storage and are therefore substandard can still make it to market. Kamagra may or may not contain sildenafil, and even if it does, the dosage may be suspect.

So, better stick to Viagra. If you get it by prescription in Canada or the U.S., you will get the right stuff. But elsewhere, for example in Mexico, you can walk into a pharmacy and buy Viagra without a prescription. And you will get a bottle with pills that look just like the authentic stuff but may be fake, produced by some clandestine manufacturer in Asia. Still, this is not a life-threatening situation. But what if you are traveling in Africa and develop an infection that requires treatment with antibiotics? Unfortunately, antibiotics are the most commonly counterfeited drugs, mostly sold in low-income nations where proper pharmaceuticals are prohibitively expensive. Now we are looking at a potentially life-threatening situation!

Fake antibiotics and fake erectile dysfunction drugs are not the only problem. Fake anti-anxiety drugs like alprazolam (Xanax), lorazepam (Ativan), and diazepam (Valium), as well as attention deficit disorder meds such as methylphenidate (Ritalin) or amphetamine (Adderall) are also available on the street and on the internet. They may look like the real thing, and their counterfeitness is only revealed with the discovery that the drug doesn't work. The COVID-19 pandemic provided new opportunities for the fraudsters who quickly came up with fake versions of hydroxychloroquine. A curious situation since the fakes worked as well as the authentic version, meaning not at all. Illicit drugs are a different case, with counterfeiting being much more prominent. Lactose powder may be used to dilute heroin, dried herbs masquerade as cannabis, methamphetamine may be sold as cocaine, and illegally manufactured "designer drugs" can contain toxic impurities.

Legitimate generic drugs have also been targeted by counterfeiters. Once the patent on a brand-name prescription drug has expired, it is perfectly legal for a manufacturer to sell a generic version as long as it has the same active ingredient and the same bioavailability as the original. Minor differences in binding agents, fillers, and colors are allowed if they have been shown not to interfere with the efficacy. Generic drugs are generally priced much lower because the manufacturer does not have to recover the cost of the research it took to bring the brand-name drug to market. In Canada, about 75 percent of all prescriptions are filled with generic drugs with very few issues arising. But generic drugs can be counterfeited just like trade-name drugs, meaning that they may contain no active ingredient, the wrong ingredient, or the wrong dose of the right ingredient.

There is yet another problem. The Canadian and U.S. generic companies are held to the same standard as the brand-name companies with every aspect of production being monitored. Products are constantly checked for purity, and immediate recalls are issued should a problem appear. However, more than half of the generics sold here are imported from India or China where they are produced in licensed plants. But just because the plants have been licensed doesn't mean they all follow best manufacturing practices. Government inspections are spotty, and the lure of taking short cuts to maximize profits is inviting. This may mean using impure solvents, unclean conditions, improper testing for bioavailability or contamination with foreign materials such as tiny shards of glass once detected in a generic version of Lipitor, the cholesterol lowering drug. Inspectors from the U.S. FDA sporadically travel to India to inspect plants producing drugs for the American market. However, these trips are announced in advance allowing the facilities to prepare and put their best face forward. Even for brand-name and generic meds produced in North America, about 80 percent of the active ingredients are imported from India or China but hopefully are checked for impurities before being formulated into the final product.

Counterfeiters and sloppy Asian generic drug manufacturers are a big problem, as the brand-name pharmaceutical industry relishes in pointing out. But there are plenty of skeletons in the brand-name closet as well. Companies have been fined for promoting off-label use of drugs and for paying kickbacks to physicians for prescribing certain drugs and for withholding known side effects. In a paper submitted in 2000 to the

New England Journal of Medicine describing studies on the painkiller Vioxx, Merck researchers failed to disclose some cases of cardiovascular events in the experimental group of which they were aware.

Of course, neither the generic nor the brand-name drug industry should be universally impugned because some cases of fraud have come to light. In North America 99 percent of prescription drugs available in our pharmacies are properly manufactured and do conform to label information. While no drug performs as well as its exuberant television ads proclaim, benefits do outweigh risks when the meds are appropriately prescribed. On the other hand, drug counterfeiters should be vigorously pursued for enticing people to play Russian roulette with their pill bottles. Sadly, we will never be able to eliminate counterfeiters. The problem has a long history, evidenced by the appearance of fake cinchona bark to treat malaria back in the seventeenth century. As long as the road of deception is paved with money, there will always be fraudsters who will take it.

SHOULD WE BE EATING MORE MUSHROOMS? MAYBE.

Scientific studies are often speckled with words like *can*, *could*, *may*, *appear*, *linked*, or *correlated*. This is bothersome. These terms are just too iffy to allow for any solid conclusions to be drawn. I wish sometimes scientists could cast doubt aside and say *will* or *does*.

Let's look at an example to illustrate what I mean. A study reported in the *Journal of Alzheimer's Disease* found "that people who integrate mushrooms into their diets 'appear' to have a lower risk of mild cognitive impairment which often precedes Alzheimer's disease." That conclusion was arrived at by researchers in Singapore upon analyzing food frequency questionnaires periodically filled out over six years by some six hundred seniors who also underwent various tests for cognitive abilities. Their summary stated that "eating more than two portions of cooked mushrooms per week (a portion being defined as three-quarters of a cup of cooked mushrooms) could lead to a fifty percent lower risk of mild cognitive impairment." They went on to say that "this correlation is surprising and encouraging." Note the *could* and *correlation*, terms that emphasize the study is permeated with uncertainty.

Correlation cannot prove a causal relationship. People who eat more mushrooms likely eat more fruits and vegetables and may have different activity levels. Consider also that food frequency questionnaires are notoriously problematic because people have difficulty remembering exactly what they ate, are not reliable in judging amounts, and are also likely to emphasize foods they think they should have eaten instead of what they actually ate. Furthermore, different varieties of mushrooms have different chemical profiles, and whether the subjects were eating oyster, shiitake, golden, or button mushrooms isn't clear. So, yes, "Eating Mushrooms Could Slash Risk of Cognitive Decline by 50 Percent" as one headline screamed. Or perhaps not.

Another recent paper in the *European Journal of Cancer Prevention* "found that eating mushrooms may reduce the risk of gastric cancer." In that review, researchers pooled a number of studies in which patients with gastric cancers had filled out food frequency questionnaires and then compared the data with that obtained from people without cancer. The relative risk of gastric cancer for the highest mushroom consumers versus the lowest was found to be 0.82. This means that high mushroom consumers have an 18 percent reduced risk of gastric cancer when compared to low mushroom consumers. That sounds significant, but there are some issues to consider.

First, "high mushroom consumption" was determined by number of servings a week, but what people consider to be a serving is variable. Furthermore, the studies differed in the number of servings that defined high consumption. Then, about the 18 percent reduction. You have to ask 18 percent of what? In other words, what is the risk of developing gastric cancer in the first place? It turns out that the incidence is about ten in a thousand over a lifetime for men, less for women. So, with an 18 percent reduction, this would mean eight in a thousand. One way to look at this is that if a thousand men ate a high mushroom diet, two would be saved from contracting stomach cancer. In other words, switching to a high mushroom diet gives you a 1 in 500 chance of avoiding stomach cancer. Again, it is unlikely that outside of their mushroom consumption, the low and high mushroom eaters have similar diets.

We are still not done. The kicker is that the mushroom benefit was seen only in Asian countries, not in Europe or America. Maybe Asians eat different varieties. Or more likely, eating mushrooms is just a marker for a different lifestyle. The bottom line here is that because mushroom eaters in some parts of the world have a lower rate of gastric cancer, we cannot conclude it is because of the mushrooms. It *may* be. Or maybe not.

The *may* can actually be strengthened by noting that mushrooms contain various compounds such as scabronine, ergothioneine, or lentinan, which in the laboratory can be shown to have nerve-growth, antioxidant, anti-inflammatory, and even anticancer effects. But that also can be said for numerous compounds found in fruits, vegetables, grains, and spices. Let's also remember that the human body is not a giant petri dish. Many compounds show activity in vitro that is not replicated in vivo.

All this being said, I think there is enough evidence to suggest that it is a good idea to incorporate mushrooms into the diet. They can serve as a tasty replacement for meat in some dishes. My favorite is mushroom goulash. Fry onions in a little extra-virgin olive oil, add sliced mushrooms (I use button, Portobello, and shiitake), a diced tomato, a chopped green pepper, garlic, pepper, and a bit of salt. If you like zucchini, you can add that as well. Then the key ingredient, a heaping spoonful of Szegedi paprika! Talk about a spice that is rich in carotenoids and polyphenols! Cook with occasional stirring until the mushrooms are soft. Goes well with spaetzli or rice. I am willing to go out on a limb and say that this dish *does* taste great, and you *will* enjoy it! It *may* even benefit your health.

THE "ELIXIR OF LIFE" OR A LOT OF BULL?

The exuberant phrase the "elixir of life" was repeated in numerous breathless accounts of a study that appeared in the prestigious journal *Science* with the alluring title "Taurine Deficiency as a Driver of Aging." Prior to the buoyant press reports, it was mostly people familiar with the list of ingredients on the label of the energy drink Red Bull who were aware of the existence of this chemical. Why does the beverage contain taurine? Actually, that is somewhat of a mystery. The only information provided on Red Bull's website is that "taurine is an amino acid, naturally occurring in the human body and present in the daily diet." No argument with that. Well, maybe a little one. Taurine is not exactly an amino acid, it is an aminosulfonic acid, and unlike common amino acids, it is not incorporated into proteins. It is found in fish and meat, and while it has many important biochemical functions, it is not an essential component of the diet because the body can make it from cysteine, an amino acid that is readily available from proteins in the diet.

It was back in 1827 that German chemists Friedrich Tiedmann and Leopold Gmelin first isolated a compound from bull bile that came to be called *taurine* from the ancient Greek word for bull or ox. A bull certainly conjures up an image of energy, which may be the reason that taurine was incorporated into the beverage. That in a curious way justifies the logo of two bulls charging at each other on every can of Red Bull despite the fact that there is no evidence of taurine being a stimulant. The stimulant in Red Bull is caffeine, though there is less present than there is in a cup of coffee.

Now on to the study. Taurine is present in human blood, and a number of conditions ranging from diabetes and obesity to hypertension and inflammation are associated with lower levels. Furthermore, blood levels tend to decrease with age. These observations do not prove that taurine has a causative role in disease or aging, but that possibility is worth exploring. That is exactly what the authors of the paper in *Science* did. Worms, mice, and rhesus monkeys fed taurine were compared in various ways with worms, mice, and monkeys who did not receive the compound. The life span of the worms and mice increased, significantly in the case of the mice. Females and males lived 10 percent and 12 percent longer respectively. Since monkeys live much longer than mice, we will have to wait for follow-up studies to see if the monkeys also live longer.

The goal of anti-aging therapies is not only to increase the life span, but also to increase health span. Extending life is not particularly desirable if the added years are filled with misery. That is why it is noteworthy that both in mice and monkeys health span was increased as determined by decreased DNA damage, reduced markers of inflammation, less fraying of chromosomes, and improved functioning of mitochondria, the organelles in cells where energy is generated. Taurine also suppressed older, damaged senescent cells that refuse to die and begin to excrete inflammatory cytokines that may trigger aging and diseases such as Alzheimer's. All that sounds great. But time to look at some numbers.

In the trials where the greatest benefits were seen, the animals were given a gram of taurine per kilogram of body weight every day. That is far, far more than the roughly 40 to 400 milligrams in our daily diet. No point reaching for a Red Bull. An average adult would have to consume about sixty-three cans a day to equal the dose of taurine given to the test animals. Neither can supplements that commonly have 500 to 1,000 milligrams per capsule mimic the amounts given to the animals. The researchers who carried out the studies state clearly that in the absence of human clinical trial, supplements are not recommended. While taurine in small doses seems safe, no humans have ever been exposed to doses of taurine equivalent to what was used in the animal studies.

Incidentally, the taurine in Red Bull is not isolated from bull semen as some circulating misinformation claims. Although the compound is found in bull semen, for commercial purposes it is synthesized in the lab and is therefore suitable for anyone wishing to avoid animal products. Another bit of misinformation is that Red Bull was the target of a class action lawsuit that claimed the beverage was falsely advertised because it did not actually "give you wings" as the slogan says. There was indeed a class action lawsuit, but it was about a lack of any evidence that the beverage had a stimulating effect beyond what a cup of coffee could deliver. Red Bull settled for $13 million, claiming that it stood by the correctness of its ads but did not want to engage in the expensive process of litigation.

The company hasn't given up on ads that portray Red Bull as being more than just a tasty beverage. A recent ad features a dog with a walking stick that after downing a Red Bull jumps on a skateboard, performs a double loop, and then flips and catches the board like a pro as a voice-over asks, "Who says you can't teach an old dog new tricks?" Cute. I wonder if the ad writers will learn a new trick from the taurine study and go on to hype Red Bull as a remedy against aging.

The fact is that aging is a complex, multifactorial process that involves numerous alterations of the myriad biochemical reactions that constitute life. Suggestions that changes in the amount of one of the thousands of biochemicals found in the bloodstream is the driver of aging, as the title of the paper in *Science* claims about taurine is . . . well, a lot of bull.

Apparently, Bryan Johnson, a self-described "professional rejuvenation athlete," doesn't think so. Johnson takes 3 grams of taurine a day along with some twenty-six other supplements! The man is quite a phenomenon. He is a multi-millionaire who made a fortune by selling the payment processing business he founded to PayPal and now devotes his life to slowing down the march of time. On interviews he often wears a tee shirt with "DON'T DIE" in bold lettering. You can buy one from his company, Blueprint, along with the supplements that he takes. The shirt only costs twenty-four bucks, but if you want to follow in Johnson's footsteps, the supplements needed can set you back hundreds a month. Although Johnson will make statements like "Death may not be inevitable," I doubt he actually thinks he can outrun the Grim Reaper. He looks on himself as an experiment to see if the biological clock can be slowed by incorporating the results of scientific studies into one's life. He jumps on any study that shows some benefit in some published trial, be it in cell culture, worm, fly, or animal. Then he takes a giant leap of faith that it may benefit humans.

Johnson, who has no scientific background, has put together a regime that he claims can reverse biologic aging and extend life. It would take too long to list all the supplements he takes, but some you may recognize include glucosamine, turmeric, cocoa flavanols, melatonin, coenzyme Q10, ashwagandha, sulforaphane, genistein, hyaluronic acid, testosterone, garlic, ginger, and taurine. Then there's N-acetylcysteine, nicotinamide riboside, tyrosine, pea protein, calcium alpha-ketoglutarate, and nordihydroguaiaretic acid. You can even add prescription drugs such as metformin, rapamycin, and 17 alpha-estradiol that have hinted at longevity. Besides supplements, Bryan follows a vegan diet, fasts for eighteen hours a day, engages in a brutal exercise regimen, and clamps a machine on his abdomen that he says produces the same effect as 20,000 sit-ups by stimulating his core muscles. He also emphasizes the importance of sleep, although I'm not sure how well he sleeps with a device strapped to his penis to monitor the extent of his night-time erections he claims is a measure of aging.

How is Johnson doing? He certainly looks ripped, and his frequent MRIs and tests for numerous biomarkers indicate that his biological age is indeed younger than his actual age. Pretty interesting, but still, this is a study of n=1, which in science no way constitutes proof.

A TALE OF TWO TWINS MEETS A TALE OF TWO CITIES

Let's start with the twins. Aimee and Nancy are identical twins who agreed to take part in an experiment organized by *Panorama*, an excellent British television documentary that often deals with scientific issues. In this case, the focus was on the impact of ultra-processed foods on health. For two weeks, Aimee consumed only ultra-processed food while Nancy followed a diet of fresh fruits, vegetables, and home-cooked meals. The meals were matched for calories, fat, sugar, and fiber so that the only difference was processing.

There is no clear definition of ultra-processed foods, but in general they come in some sort of package and are constructed from a number of food components that are blended, usually using some sort of machinery, with sugar, fat, salt, and a host of additives that can include preservatives, sweeteners, emulsifiers, humectants, artificial flavors, and artificial colors. Breakfast cereals, frozen pizza, instant soups, sliced white supermarket bread, hot dogs, potato chips, and, alas, Montreal smoked meat, are typical examples. There is no hard and fast rule, but a food that comes in a package and has a long list of ingredients including some like hydrolyzed protein, modified starch, high-fructose corn syrup, interesterified oils, and hydrogenated fats that would not be found in a home kitchen can be classified as ultra-processed.

Before and after the *Panorama* experiment, Aimee and Nancy underwent a battery of tests monitored by Professor Tim Spector of King's College. The results were surprising. Aimee gained a kilo, Nancy lost weight. Aimee's cholesterol and blood glucose went up, Nancy's went down. And that was after just two weeks! What is it about processed food that is responsible for this nefarious effect? Usually processed foods are castigated for their fat, sugar, and salt content, but in this case, the two diets were matched in terms of these components.

Could it be a difference in texture? Studies have shown, for example, that powdered oats lead to a greater spike in blood glucose than steel-cut oats, presumably because the processing breaks down cells and leads to enhanced absorption of carbohydrates. Can it be the packaging? Phthalates, perfluoro alkyl substances (PFAS), and bisphenol A, all of which can be found in plastic or paper packaging, have been blamed for all sorts of health effects. Or is it that ultra-processed foods tend to be consumed faster so that hunger satisfaction signals do not kick in and more food is consumed? It seems "fast food" applies not only to the speed with which it is served, but also to the speed with which it is downed.

Then there is the question of the additives. Emulsifiers are one class of additives that are being scrutinized because they are ubiquitous in processed foods. They stop oil and water from separating, they make ice cream smooth, prevent bread from crumbling, and slow the formation of "white bloom" on chocolates which is the result of fat separating out. But they may be doing something else as well. Emulsifiers may disrupt our microbiome, meaning that they alter the balance of bacteria in the gut. An unhealthy microbiome has been linked with all sorts of health problems ranging from diabetes and obesity to depression and inflammatory bowel disease. The incidence of the latter, particularly in the form of Crohn's disease and colitis, has been increasing in recent years, in step with the increase in processed food consumption. In studies with mice, two popular emulsifiers, carboxymethyl cellulose (CMC) and polysorbate eighty, have been shown to cause dramatic changes in gut bacteria in favor of pro-inflammatory species. Furthermore, the mucus lining of the animals' gut thinned, allowing bacteria to invade and inflame the gut wall. Human studies are needed, but these present challenges. There are over sixty different emulsifiers in the food supply, and they have different properties. Some may be injurious to the gut, some totally safe. And a randomized study would require one group of subjects consuming a totally emulsifier-free diet, which is almost impossible since thousands of foods contain emulsifiers. However, it is possible to have people fill out dietary questionnaires from which emulsifier intake can be estimated. They can then be followed for years to see what disease patterns emerge.

And that takes us to the tale of two cities. In Paris, a study of over 100,000 adults coordinated by the Sorbonne's Dr. Mathilde Touvier did just that and found that exposure to emulsifiers was associated with an increased risk of cancer. Of course, there is no nail in the coffin here, since it is possible that emulsifiers are just a marker for processed food consumption in general. Indeed, Dr. Touvier also found an association between ultra-processed foods and cancer. Now over to London. The U.K. Biobank study, coordinated by researchers at Imperial College

London, followed close to 200,000 people for thirteen years and also found an association between eating ultra-processed foods and cancer. Both studies controlled for physical activity, body mass index, calorie intake, alcohol consumption, and smoking. How strong was the association? The hazard ratios are around 1.1, meaning that over the period of the study, people who ate the most processed foods compared with those who ate the least had a 10 percent greater risk of being diagnosed with cancer.

The usual admonition that correlation is not the same as causation applies here. But with the potential detrimental effects of ultra-processed foods on health, there is just too much smoke for there to be no fire. So, tonight, no Montreal smoked meat for me. Although it has no emulsifiers, it does have salt, sugar, nitrites, sodium phosphate, irradiated dehydrated garlic, soy flavor, and sodium erythorbate. That means it qualifies as an ultra-processed food. But I won't swear off it forever. Tastes too good.

IS HEALING WITH WATER ALL WET?

"Oh, you won't be drinking it" crooned Anthony Hopkins as Dr. John Harvey Kellogg in the 1994 film "The Road to Wellville." The movie takes us back to the late nineteenth century and tells the story of the Battle Creek Sanitarium, America's largest ever "wellness institute," where Dr. Kellogg catered to the needs of the rich and famous with a plethora of unusual treatments. The likes of Thomas Edison, Henry Ford, and Mary Todd Lincoln came to "take the cure." They sat in a lightbox, ate a vegetarian diet, exercised, and listened to Kellogg's lectures about the evils of sexual activity. But mostly they were subjected to hydrotherapy. They sat in hot and cold baths, were hosed with cold water, and braved being wrapped in wet sheets.

For the tour de force, they endured water jetted into their colon since Dr. Kellogg believed that to be the part of the anatomy where all disease begins. A thorough cleansing would put patients on the path to health! Sometimes Dr. Kellogg determined that a more advanced form of therapy was needed and called for his enema machine to be loaded with yogurt instead of water. Hence his remark to the patient who stated his dislike for yogurt upon hearing Dr. Kellogg give instructions to an attendant to "bring in the yogurt."

Actually, the yogurt treatment was not total nonsense. Kellogg was a devotee of Ilya Metchnikoff, the Russian Nobel Prize winner who had advocated for the use of lactic acid bacteria, as found in yogurt, for a healthy and long life. Dr. Kellogg's yogurt enemas can be considered to be an early application of probiotics, the introduction of "good" bacteria that proliferate to crowd out the "bad" varieties that cause disease.

While not all patients encountered the delights of yogurt in this fashion, they were all subjected to some sort of water therapy. That, unlike breakfast cereals, was not a Kellogg invention. The ancient Greeks believed that certain natural springs were blessed by the gods to cure disease. Hippocrates, who eschewed such myths, recommended bathing in mineral water to restore the body's balance of "humors." The Romans used hot thermal waters to relieve the pain of arthritis, and the epithet "baths, wine, and sex make life worth living" can be found on some Roman tombstones.

Hot springs maintained their therapeutic reputation throughout history as evidenced by Mark Twain's comment after finding that his rheumatism was soothed by bathing in the spring waters of Aix-les-Bains in France. He claimed the experience was "so enjoyable that if I hadn't had a disease, I would have borrowed one just to have a pretext for going on."

English physician Sir John Floyer was the first to investigate water as therapy in 1702. He was intrigued by peasants in Lichfield bathing in cold water to treat their various ailments and became convinced that cold water therapy was effective. Floyer managed to persuade "worthy and obliging gentlemen" to contribute towards building a cold-water bath in Lichfield that was supplied by water from the spring. It turned out that his patients were not as obliging when it came to immersing themselves in the freezing water. Dr. James Currie of Liverpool had somewhat greater success with his 1797 pamphlet "Medical Reports on the Effects of Water, Cold and Warm, as a Remedy in Fevers and Other Diseases," but it was Vincenz Priessnitz, born in 1799 in Grafenberg, now in Czechia, who put hydrotherapy on the map.

The story is that as a youngster Priessnitz watched a wounded deer bathe in a pond and later saw the same deer with its wound healed. When as a teenager he was run over by a horse cart, he remembered the deer and wrapped himself in wet bandages to treat his broken ribs. Soon the pain was gone, and the ribs healed, despite the local physician having told him that the injury would forever be crippling. The story of the miraculous treatment spread, and soon people began to approach Priessnitz for advice about treating their ailments. This usually involved being wrapped in wet cloths and lying under blankets until they perspired profusely. Then it was time for a plunge into cold water. Priessnitz's rationale was that the quick change in temperature opens the pores in the skin and allows unnamed "bad substances" in the blood to escape. There was about as much evidence for this as for any of today's "detox" regimens.

Nevertheless, claims of cures abounded. In 1822, catering to the enhanced demand for his services, Priessnitz expanded his father's farmhouse to serve as a spa where patients received treatment including cold showers with streams so strong that some complained of "being flattened" by them. Apparently, Captain Richard Claridge was not one of the complainers. Quite the opposite. Suffering from headaches and rheumatism, he was so satisfied with his treatment at Grafenberg that in 1842 he published the book *Hydropathy, or The Cold-Water Cure, as Practiced by Vincent Priessnitz* and toured the U.K. with lectures singing the praises of this wonderful treatment. Soon water cure establishments sprouted up all over Britain although they did not escape criticism. One clever writer claimed to have watched a duck fluttering about in a cold pond expecting it to shout, "Priessnitz forever" but only crying out "Quack! Quack!"

A major boost to hydrotherapy was given by German Catholic priest Sebastian Kneipp, who in 1847 fell ill with tuberculosis and having heard of water therapy took to bathing in the cold Danube River. He recovered and became a devotee, although certainly the cure was not due to cold water. Water does not kill the bacteria that cause tuberculosis! Still, personal experience can be very convincing, and Kneipp launched a career as a healer, expanding on Priessnitz's regimen by incorporating botanicals, exercise, and a simple diet based mostly on plants and whole grains. In 1891, Kneipp founded a company to market botanicals. It still exists today, offering "therapeutic" bath products such as "relaxing lavender mineral bath salt" and "joint and muscle arnica mineral bath salt." Claims of "proven" effectiveness should be taken with a grain of salt.

Benedict Lust, who also believed to have been cured of tuberculosis by Kneipp's method, established a Kneipp Water Cure Institute in New Jersey and blended hydrotherapy with his opposition to processed foods and prescription drugs. Believing in the inherent ability of the body to heal itself, Lust is regarded as the father of naturopathy in America.

In his 1901 book, *Rational Hydrotherapy*, Dr. Kellogg recounted the stories of his forerunners and claimed to have improved on their methods by showering the body on the inside as well as the outside. He was on firmer footing when he drenched his patients with lectures about the evils of smoking, drinking, and eating the standard fare of salted meats, white bread, and fried foods.

PROCESSED FOODS HAVE MANY FACES

The cruise ship had a restaurant dedicated to molecular gastronomy, so of course I was eager to try it. I was familiar with the term. It had been introduced just a few years earlier in 1988 by French chemist Hervé This and Hungarian physicist Nicholas Kurti to describe a discipline that "seeks to generate new knowledge on the basis of the chemistry and physics behind culinary processes."

The duo was not the first to explore the science of cooking. As early as 1825, French lawyer Anthelme Brillat-Savarin composed *The Physiology of Taste*, described as a gathering of "the intelligent knowledge of whatever concerns man's nourishment." In 1847, Justus von Liebig, one of the greatest organic chemists of the nineteenth century, published *Researches on the Chemistry of Food* in which he, as it later turned out erroneously, argued that cooks should sear meat to retain its fluids. Then in 1984, Harold McGee published his masterpiece *On Food and Cooking: The Science and Lore of the Kitchen* in which he discusses in detail where our foods come from, their chemical composition, and how cooking changes their chemistry.

While I have thumbed McGee's book until the pages have become ragged, my interest in the chemistry of cooking traces to another book published a year earlier: *The Cookbook Decoder, or Culinary Alchemy Explained* by Arthur Grosser. Dr. Grosser had been my physical chemistry professor at McGill, so of course I was interested in reading his book. In a witty fashion he talks about puncturing an eggshell with a pin to prevent the horror of a hard-boiled egg with a flat bottom, and why cloves of garlic should be mashed before adding to a dish. That's because two chemicals have to combine to create allicin, the compound responsible for the flavor of garlic, which only happens when cell membranes break.

Hervé This and Kurti went one step further than just discussing the chemistry of cooking, they introduced the idea of molecular cuisine to explore novel culinary methods rooted in science. It was a couple of those "novel methods" I experienced dining in the cruise ship's avant-garde restaurant. The first item, served in a test tube, consisted of beads that popped in the mouth to release an orange flavor. This was an example of spherification, one of the most common techniques in "molecular cookery." From having read Hervé This's book, I knew the chemistry involved. When orange juice is mixed with calcium chloride and added one drop at a time to a solution of sodium alginate extracted from brown seaweed, a thin membrane of calcium alginate forms trapping the orange juice in pearl-like beads.

This "appetizer" was followed by a soup that was essentially a flavored foam. I'm not sure how that was prepared, but the usual method is to whip a liquid, in this case a soup base, with a hand mixer until it forms a foam that is then stabilized by the addition of the emulsifier lecithin extracted from soybeans. The foam dissolves in the mouth bathing the taste buds in flavor before it vanishes into thin air.

All in all, the encounter with chemistry in this setting was enchanting but the gustatory merit of the experience was questionable. However, what was not questionable, was that the meal was composed of highly processed foods! And that is what now sparked this decades-old recollection of my fling with molecular gastronomy. Today, the scientific and popular literature scream about how ultra-processed foods are wreaking havoc with our health. But like just about every nutritional issue, the science is nuanced.

Unless we are talking about an apple picked off a tree, or a tomato plucked off a vine, our diet consists of processed foods. Cooking is a process, as is pickling, drying, or freezing. Blending oil with vinegar and adding egg yolk as an emulsifier to make mayonnaise is a process, as is tenderizing a steak with pineapple juice. But what do we mean by ultra-processed foods?

A simple definition is elusive, but akin to pornography, you know it when you see it. A common description is that a food is deemed to be ultra-processed if it has more than one ingredient that you rarely find in a kitchen and is commonly packaged in a container that has a label with a list of ingredients such as preservatives, emulsifiers, sweeteners, artificial colors, and flavors that are not typically used in home cooking. Also, ultra-processing may include high pressure hydrogenation of fats, hydrolyzing vegetable proteins with acids, and the use of extrusion equipment to produce cereals. Cold cuts, potato chips, breakfast cereals, frozen pizzas, instant soups, commercial ice cream, flavored yogurt, plant-based meat substitutes, mass produced packaged bread, chicken nuggets, and french fries would fall into the ultra-processed category.

Roughly 60 percent of the Western diet is made up of ultra-processed foods. So what? Unfortunately, numerous studies have linked the consumption of these foods with an increased risk of cardiovascular disease, cancer, type 2 diabetes, and obesity. Although an association cannot prove a cause-and-effect relationship, attention is warranted when so many studies all point in the same direction. The prevailing opinion has been that the negative effects of ultra-processed foods are due to their high content of sugar, fat, and salt. However, an epic clinical trial by Dr. Kevin Hall of the National Institute of Diabetes and Digestive and Kidney Diseases that had subjects eat ultra-processed food for two weeks and then unprocessed food for two weeks, revealed that there was more to the issue. The participants were allowed to eat as much as they liked of the meals, both of which contained the same amount of sugar, salt, fiber and fat. Yet the subjects on the ultra-processed diet ate about five hundred calories a day more and put on about 0.9 kg while the other group lost about the same amount. Since both diets were judged to be equally palatable, some other factor is involved.

One possibility is that ultra-processed foods have their structure altered by whatever processing they undergo making them softer so that they can be consumed faster. More calories are consumed per minute, meaning that a feeling of fullness is delayed until long after the food has been consumed. Protein leverage is another possible issue. We seem to have a biological need to consume a certain amount of protein a day and ultra-processed foods are generally lower in protein, prompting greater consumption.

As far as a link to disease, additives such as emulsifiers and sweeteners may change the composition of the gut microbiome, reducing the bacteria that produce short chain fatty acids that can keep inflammation in check. For whatever reason, it is becoming more and more clear that ultra-processed foods have a negative effect of health and that home cooking is the way to go.

Finally, I don't think I was in any way harmed by the lecithin-laden spherical beads served in a test tube, but as much as I like chemistry, I don't think you need chemical gimmicks to serve good food.

EATING FRENCH FRIES IS NOT THE SAME AS SMOKING CIGARETTES

I had never heard of psychiatrist Dr. Paul Saladino which is somewhat surprising because he is quite frisky in the duck pond. His TikTok videos in which he tries to convince his legions of followers that dietary fiber is unnecessary, that drinking beer leads to "manboobs," that LDL cholesterol does not increase the risk of heart disease, that oatmeal is toxic, and the key to health is eating red meat, are laughable.

Saladino's pseudoscientific rants were brought to my attention by a former student who now teaches science in Germany. He was asked by one of his students about a video in which Saladino claims that eating a serving of McDonald's fries is equivalent to smoking a pack of twenty-five cigarettes.

The stimulus for this video seems to be a paper that Saladino read but was unable to properly digest. It discussed similarities between the chemical content of french fries and tobacco smoke and noted that a serving of fries can contain some carcinogenic aldehydes in amounts comparable to that found in the smoke from twenty-five cigarettes. In no way did the authors suggest that the risks were comparable.

Let's note right away that there is a big difference between inhaling or ingesting a substance. Inhalation leads to direct entry into the bloodstream while the digestive tract contains numerous enzymes that metabolize food components. Next, tobacco smoke contains literally thousands of compounds, with sixty-two of these listed by the

International Agency for Research on Cancer (IARC) as carcinogenic to humans. The most significant carcinogens in tobacco smoke are not aldehydes but N-nitrosamines, polyaromatic hydrocarbons, aromatic amines, 1,3-butadiene, benzene, and ethylene oxide. While there is no question that carcinogenic aldehydes such as crotonaldehyde can form when fats are heated, the total number of carcinogens that invade a body from a pack of cigarettes is far, far greater than from a serving of french fries.

Of course, the only way to compare the health impact of a daily serving of fries to smoking a pack a day would be to run a long-term study comparing two groups of subjects with the only difference between them being smoking or french fry eating. Clearly this is impossible to do, but if it were carried out, I would wager that the smoker group would have a far higher incidence of cancer than the french fry group.

Fearmongering has become an industry, and Saladino is a head honcho in this area. The usual technique is to pick a scientific study that finds some risk, and then exaggerate it without taking into account type and extent of exposure. Pesticides, fluoride, oxalates, gluten, lectins, and vaccines have all been unrealistically portrayed as villains. This is not to say that there are no legitimate chemical risks. We live in a very complex world, with some 160 million known chemicals, both natural and synthetic. There certainly are issues with some of these. Perfluoroalkyl substances (PFAS), bisphenol A, and phthalates are present in just about everyone's bloodstream and may indeed be causing some serious mischief.

One way or another, we are in contact with thousands of chemicals on a regular basis and teasing out individual effects is not possible. While french fries may indeed contain some carcinogens, it does not automatically follow that eating them causes cancer. As a classic analogy, coffee contains a number of carcinogens such as furfural, caffeic acid, and styrene, but we know that coffee doesn't cause cancer.

None of this is to say that I am willing to absolve french fries from all blame. Excessive consumption of fried foods is a problem, and not only because of the extra calories provided by the fat. When fats are heated, particularly polyunsaturated seed oils, they form a slew of potentially carcinogenic compounds. And then there is the issue of the Maillard reaction, named after Louis Camille Maillard, physician turned chemist, who in 1912 described the reaction between sugars and amino acids that produces a variety of melanoidins, which are responsible for the browning of toast, doughnuts, and french fries. In fried potatoes, glucose and the amino acid asparagine undergo a Maillard reaction to yield acrylamide, classified by IARC as a probable human carcinogen.

Although associations cannot prove a cause-and-effect relationship, a study by the highly reputable Fred Hutchinson Cancer Research Center in Seattle compared about 1,500 prostate cancer patients with the same number of controls and found that regular consumption, at least once a week, of fried chicken, fried fish, doughnuts, or french fries increases the risk of developing the disease.

While carcinogens in fried foods cannot be totally eliminated, they can be significantly reduced. The secret is to do your frying in an air fryer. These devices have taken kitchens by storm, including mine. Basically, they are small convection ovens in which a current passing through an element heats air that is then circulated by a fan. The basket in which the food is placed has openings to ensure heating from all sides, so covering these with parchment paper or aluminum foil in pursuit of cleanliness is counterproductive.

Although the temperature to which the air is heated, about 180 to 190 degrees Celsius, is comparable to the temperature of frying oil, air is far less efficient at transferring heat to food. While deep frying takes only five or six minutes, air frying can take three times as long. However, since no oil is being used, there is no worry about its carcinogenic breakdown products. Furthermore, hot air penetrates the food less effectively than hot oil, so the inside of the food doesn't get as hot which means significantly less acrylamide formation.

As far as crispiness goes, that is determined by the moisture content at the food's surface. When food is placed in a deep fryer, the immediate bubbling seen is due to steam released from its surface. Hot air does not heat the surface quite as well, but still well enough to drive out moisture and produce crispiness. In the case of fries, this can be improved by first coating the potatoes with a thin layer of oil. If you really want to reduce oil degradation products, the best choice is avocado oil because of its extremely high smoke point. I won't say that my "air fries" are comparable to the best double-fried restaurant version, but they are very acceptable. And healthier.

Remember that the claim of french fries being as dangerous as smoking comes from someone who thinks that lamb testicles and raw liver are healthy, and cruciferous vegetables like broccoli, brussels sprouts, chard, and kale are "bullshit." These, Saladino says, should be avoided, because "once chewed they produce sulforaphane which is toxic to humans." Actually, sulforaphane has been shown to be an anti-carcinogen. So go for your broccoli and kale. If it is taste and crispiness you are after, put them in the air fryer. As far as Saladino's TikTok videos go, after watching a bunch of them with their confusing message, I am led to conclude that this psychiatrist needs a psychiatrist.

THE NEBULOUS POSITIVE EFFECTS OF NEGATIVE IONS

I've always enjoyed visiting Niagara Falls. Watching that mass of water hurl over the edge and crash onto the rocks below is an awesome sight. But it may be more than that. Breathing the air around the falls may have a positive effect on health. That's due to its content of negative ions, or "air vitamins," as they are called by some overly enthusiastic supporters of their potential health benefits.

Our story starts with a discovery made by the 1905 Nobel Laureate in physics, Philipp Lenard. The man was undoubtedly a genius. And a despicable human being! Educated in Hungary, then in Germany under famed chemistry professor Robert Bunsen, Lenard became an early member of the Nazi party and a huge supporter of Hitler who made him head of "Aryan Science." Lenard was a fervent antisemite, who in a speech once proclaimed that "the Jews must be sunk right down to the center of the earth." "Germany should rely on German physics" he maintained, "not on the Jewish fraud" as perpetrated by Einstein whom he viciously attacked for having introduced a "Jewish spirit" into physics. Lenard also resented Wilhelm Roentgen's being awarded the 1901 Nobel Prize in physics for the discovery of X-rays, a discovery he claimed to have made.

While Lenard's politics and character were contemptible, his science is deserving of accolades. Of interest here is his 1892 paper in which he described how the splashing of falling water charges the surrounding air with electricity, known today as the Lenard effect. At the heart of this effect are electrons, the "glue" that bind atoms together in a molecule. These bonds can be broken by the input of energy, such as water smashing into a surface. On impact, some water molecules are wrenched apart and the negatively charged electrons that had held the atoms together are released, only to be snatched up by oxygen, nitrogen, and carbon dioxide, components of the surrounding air. These now become negatively charged and constitute the negative ions that can be detected around waterfalls.

Prompted by Lenard's discovery, a number of scientists became interested in the effects of breathing such "electrified air." Although these studies were generally not of high quality, they seemed to agree that negative ions can produce a positive effect. In the 1950s, Professor Albert Krueger, a University of California bacteriologist, was the first scientist to take a serious interest in the effects of negative ions on living tissues. He chose to study the trachea, or windpipe, of animals, since it is the primary zone of contact with inhaled air ions. By the time Dr. Krueger began his experiments, it was known that air could also be ionized by radioactive substances. He used beta radiation, essentially a barrage of electrons, emanating from radioactive tritium to generate negative ions that were then directed at trachea tissue. Krueger was able to document a variety of biochemical changes, including alterations in the production of neurotransmitters such as serotonin.

The demonstration that inhalation of negative ions can have an effect on biology unleashed a cavalcade of research. Plants, it was found, also release negative ions as they photosynthesize. This was quickly linked with the supposed benefits of "forest bathing." A number of studies, albeit not ones I would call compelling, have indeed suggested that spending time in green spaces results in enhanced relaxation, reduced tensions, and an improved mood. This was now rationalized on the basis of trees producing negative ions.

If those trees surround a waterfall, the beneficial effects might even be greater! So, Chinese researchers enlisted volunteers, including some suffering from chronic fatigue syndrome, to spend a week in the vicinity of the Huangguoshu Waterfall, the largest in Asia. The subjects filled out questionnaires on the basis of which the extent of their fatigue, anxiety, and depression were evaluated. Their scores improved as the week passed. Furthermore, blood tests showed a decrease in markers of inflammation and an increase in the blood's antioxidant capacity.

None of this is what we would call hard science, but it is enough grist for the mill that grinds out pseudoscience. Like the ability of Himalayan salt lamps to improve health. These lamps, made by placing a lightbulb inside a block of Himalayan salt, are claimed to release negative ions. Identifying the salt as "Himalayan," even though it is mined in Pakistan and not in the Himalayas, adds a certain new-age mystique. But mystique is all that the lamp provides. No negative ions in sight, as can be shown by placing the salt lamp near the intake port of a mass spectrometer, an instrument that can detect even the presence of fleeting ions.

At least salt lamps pose no danger and can even look attractive. The same cannot be said for jewelry that is made of tourmaline, a semi-precious stone claimed to release negative ions. The hype is that wearing these bracelets, pendants or rings will relieve pain, increase circulation, and enhance mood. Never mind that negative ions have not been shown to relieve pain or increase circulation, tourmaline is not radioactive, so cannot generate any negative ions. Yet, when some of these items were tested, they proved to be radioactive, and therefore theoretically capable of generating negative ions. It turns out that such items had been doped with radioactive thorium oxide to an extent that if worn constantly for a year, the wearer would be exposed to a dose of radiation that exceeds the annual recommended limit. Amazingly, some of these items are claimed to offer protection against the nonexistent risk of electromagnetic radiation from cell phones by exposing the duped to the real risks of radioactive thorium. Touted as "a perfect gift for anyone who cares about their health," they are anything but.

While the positive effect of negative ions on health is somewhat nebulous, their ability to reduce particulate matter in the air is on a sounder footing. Ion air purifiers use high voltage to induce a corona discharge that generates negative ions that then attach to dust particles giving them a negative charge. These are then attracted to positively charged plates in the machine, or to walls, TV screens, and furniture that tend to have a positive charge. Unfortunately, many of these air purifiers produce unacceptable levels of ozone, and those that don't are not very efficient.

As far as testing out the effects of negative ions on mood, Niagara Falls seems to be the place to go. Last summer I had a chance to take a ride on the Maid of the Mist, the boat that approaches the bottom of the falls closer than one would think possible. If there is a place to be bathed in negative ions, this is it. I waited for the mood enhancing effect to set in. All I got was . . . wet.

DEBUNKING THE DIRTY DOZEN

It is spring and that means a flood of questions come my way about pesticide residue on produce. That's because it is at this time of the year the Environmental Working Group (EWG), a nonprofit consumer advocacy organization in Washington, D.C., annually freaks out the public with a press release that unveils the fruits and vegetables with the highest levels of pesticide residues.

The "Dirty Dozen" is the name EWG has coined for its targets, pilfered from the 1967 movie by that title. This year, the ignominious pack is led by the strawberry, with spinach nipping at its heel. EWG does an excellent job garnering publicity for its Dirty Dozen, implying anyone concerned about health should choose organic versions of the condemned fruits and veggies. EWG presents itself as a knight in shining armor, a sentinel ready to slay all those dangerous products spewed out by companies that put profits ahead of safety.

The first thing I noted when I checked the list of the Dirty Dozen is the total absence of numbers. Curious, considering numbers are the currency when it comes to discussing toxicity. Actually, the absence of data was the second thing I noted. The first item that appeared when I Googled EWG's list was a request for a donation so that these guardians of public health can continue to carry out their stated mission of "empowering consumers with breakthrough research to make informed choices and live a healthy life in a healthy environment." However, sometimes those "informed choices" may be misinformed. Also, EWG's claim of "breakthrough research" is a bit of a stretch. To arrive at the Dirty Dozen, they used information gleaned from the U.S. Department of Agriculture's annual testing of a random selection of fruits and vegetables for pesticide residues.

It is important to have a look at how pesticides are regulated. And they are, thoroughly. In Canada, the task falls to the Pesticide Management Regulatory Agency (PMRA), while in the U.S., responsibility lies with the Department of Agriculture (USDA) and the Food and Drug Administration (FDA). These bodies evaluate laboratory data, animal studies, and potential human exposure to determine a risk/benefit ratio for each pesticide. A substance is allowed if its benefits are found to outweigh risks, but that does not mean free rein of use. How and when a chemical can be applied is carefully regulated, as are tolerable residues.

A great deal of attention is paid to safety since pesticides are inherently toxic. After all, they are designed to kill insects, fungi, and weeds.

Residue tolerance levels, also referred to as *maximum residue limits* (MRL), are determined by evaluating animal studies, chemical properties such as estrogenic effects, occupational exposure, and non-occupational exposure through diet and drinking water. It is common to determine a *no observed adverse effect level* (NOAEL), which is the maximum amount that produces no effect in an animal. This is then divided by a safety factor of 100 to establish an MRL. Of course, no regulatory system is perfect, since so-called "cocktail effects" that may result from various combinations of chemicals, or subtle effects that may turn up after decades of exposure, are virtually impossible to predict.

So, how do tolerance levels stack up? Studies of produce randomly purchased both in North America and Europe consistently show only 2 to 3 percent exceed the MRL (remember this has a 100-fold safety factor built-in) and roughly half have no detectable residue at all. But even within tolerance levels, produce can harbor more or less residue, which is the basis for EWG's Dirty Dozen. Without any numbers, without any specification about whether produce in the Dirty Dozen exceeds maximum residue levels, the list is meaningless and creates unnecessary fear about eating fruits and vegetables.

This is not a trivial concern, since evidence linking fruit and vegetable consumption to health is overwhelming. Anyone who steers away from eating foods that appear on the Dirty Dozen list is doing themselves a disservice. As far as nutritional differences between organic and conventional produce go, studies can be dredged up to favor either one. Whatever difference there is may be too small to change the diet's effect on health. Where organic may have an advantage is in its reduced effect on the environment. To give EWG some credit, they do urge people to eat more fruits and veggies be they conventional or organic, but their whole approach casts a shadow on conventionally grown produce.

It is interesting to note most of the residue detected are fungicides to control the growth of mold that can produce potentially dangerous mycotoxins such as the carcinogenic ochratoxin A. Some studies have found higher concentrations of mycotoxins in organic produce since it is not protected by fungicides.

EWG's Dirty Dozen lists only conventionally grown produce and implies organic foods are grown without pesticides. Indeed, EWG urges consumers to eat organic when possible for that reason. But the message pesticides are not used in organic agriculture is misinformed. There are many pesticides allowed, with the stipulation they come from a natural source. Copper sulfate, lime sulfur, neem oil, pyrethrins from chrysanthemums, and spinosyn from a soil bacterium can all be used in organic agriculture and can, like any other pesticide, leave a residue.

Keep in mind that whether a chemical is natural or synthetic has no bearing on its potential toxicity. The reason there is no need to be concerned about organic pesticides is they are regulated the same way as any other pesticide. As far as establishing safety, regulatory agencies make no distinction between conventional and organic pesticides. Both have to jump through the same hoops and over the same hurdles.

That being said, there is no doubt a greater variety of residue can be found on conventional produce since there are far more pesticides approved for conventional than for organic agriculture. However, it is not the number of different pesticide residues that matters, but whether they exceed maximum residue limits. And, unless EWG can show that to be the case, its Dirty Dozen list amounts to no more than a ploy to raise funds.

SORDID MEDICINE SHOWS EXPLOITED INDIGENOUS CURES

Dr. Quinn, Medicine Woman, heroine of the 1990s television western by that name, routinely prescribed willow bark tea for just about any ailment, a practice she learned from the Cheyenne. Historical accounts and archaeological evidence indicate that Indigenous people in North

America really did use willow bark to treat aches, pains, fevers, and other ailments long before eighteenth century cleric Reverend Edward Stone introduced willow bark as a medicine in England.

Apparently, Stone once chewed on the bark after observing that the local people were using it to alleviate pain and fever. He found it to be bitter, and knowing that cinchona bark, which at the time was used to treat malaria, also tasted bitter, he thought that maybe willow bark could also be used as medicine. Reverend Stone had a chance to give it a shot when he developed a fever. It worked! He began to recommend it to his parishioners, and in 1763 he reported his results to the Royal Society. Today we know that white willow bark contains salicin, a compound that does actually have fever- and pain-reducing properties but irritates the stomach. In 1898, chemist Felix Hoffmann, under the direction of Arthur Eichengrun at the F. Bayer Co. converted salicin to acetylsalicylic acid and sold it as Aspirin.

Native Americans have a long history of using plants to treat ailments, a tradition that is shared with ancient Greek and Egyptian medicine, traditional Chinese medicine, and Ayurvedic medicine. One of the first documented uses of Indigenous medicine in North America was by French explorer Jacques Cartier whose crew was decimated by scurvy during the winter of 1536 spent at Stadacona, now Quebec City. A number of the men were cured by a decoction made by boiling the leaves and bark of an evergreen tree that Cartier learned about from the Iroquois. Exactly what that tree was remains a mystery, but pine needles are known to contain vitamin C, a deficiency of which is the cause of scurvy.

Numerous other plants, such as black cohosh, echinacea, sarsaparilla, golden seal, skullcap, witch hazel, mugwort, yarrow, sweetgrass, sage, cranberry, saw palmetto, ginseng, and garlic were used by American natives. Undoubtedly, some of these were useful. Black cohosh contains estrogen-like compounds that can relieve hot flashes, sweetgrass has the "blood thinner" coumarin, and ginsenosides in ginseng may help with some digestive problems.

For white American settlers, "Indians" (a term not controversial at the time) held a certain fascination. To some they were "savages," while to others they were clever children of nature who possessed secrets of healing unknown to the white man. It was that belief on which two colorful characters, John Healy and Charles Bigelow, capitalized in the late 1800s. Traveling medicine shows that featured a variety of minstrels, burlesque acts, magicians, jugglers, fire-eaters, and comedians flourished at the time. The acts drew crowds who were then subjected to a sales pitch for some patent medicine from the "Professor" or "Doctor," who was usually neither. Healy and Bigelow had the idea of taking advantage of the mystique surrounding Indigenous people to hype their version of a miracle cure they called "Sagwa."

In 1881, they founded the Kickapoo Indian Medicine Company and came up with a story to captivate their audience. Bigelow claimed he had been acting as a government scout in "Indian territory" when he came down with a fever that put him at death's door. He would not have survived, the story went, had he not been given a remedy by a Kickapoo native medicine man. Snatched from the jaws of death by this concoction, he prevailed upon the medicine man to divulge its secret ingredients, and Sagwa was born. A picture of a Native American dominated the bottle's label, underlining the message that the mysteries of natural cures that had armed natives with an iron constitution were now being revealed to the white man.

Bigelow and Healy hired Indigenous people, none of whom were Kickapoos, to perform in their medicine shows. Teepees were set up around the stage, powwows were mimicked, war dances performed, tomahawks waved, and naïve versions of "cowboy and Indian conflicts" acted out. Then, to the fierce beating of tom-toms, a "native elder" would be introduced by a name such as Thunder Cloud or Dove Wing to praise the wondrous native cures in their own tongue. The pitch doctor interpreted what was supposedly said prompting a buying frenzy as assistants circulated in the crowd with bottles of Sagwa, occasionally yelling out "all sold out, Doctor."

Sagwa, which was promoted as a "blood, liver, stomach, and kidney renovator," listed capsicum, mandrake, guaicum, sal soda (sodium carbonate), and alcohol as ingredients. Curiously, none of these were actually native remedies, but the inclusion of alcohol undoubtedly increased appeal. The Kickapoo Company even secured an endorsement from "Buffalo Bill" Cody who actually had a history of fighting Native Americans. He had his own Wild West Show, a decidedly racist endeavor in which his legendary battles were re-enacted, painting a picture of natives as savages who were conquered by the superior white men. Nevertheless, Cody gave the natives credit for Sagwa, declaring that "Kickapoo Indian Sagwa is the only remedy the Indians ever use, and has been known to them for ages." He went on, "An Indian would as soon be without his horse, gun, or blanket as without Sagwa." The truth of course was that all Native Americans were without Sagwa, which was the invention of two white men who had no connection to the Kickapoos and had founded a company with a fabricated history that hyped a concoction that had no evidence of any efficacy.

The success of the Kickapoo medicine show spawned a host of imitators peddling copies such as "Awaga" and "Sagwah," but all medicine shows began to lose their oomph with the passage of the 1906 Pure Food and Drug Act in the U.S. that prevented the sale of "adulterated products," a category that included most patent medicines since they included unapproved ingredients such as alcohol and opiates.

Unfortunately, the real history of Indigenous medicines is clouded by the sordid history of the medicine shows that exploited native culture. Recently though, the Canadian government has called for a recognition of the contributions of Indigenous science, and a "bridging, braiding, and weaving with Western science." That is a noble thought, but science is a matter not of philosophy but evidence. There is no Western science or Indigenous science. There is only science. Evidence needs to rule.

Incidentally, you can still buy white willow bark powder on the internet, often with references to ancient native wisdom. But I think today even Dr. Quinn would prefer aspirin.

IS IT EVIDENCE-BASED OR EMINENCE-BASED?

There is no question that Linus Pauling was an eminent chemist, given that "eminence" is defined as fame or recognized superiority within a particular sphere or profession. In this case, chemistry. Pauling was a Nobel Laureate and published an incredible 1,200 papers and books, including a textbook on chemical bonding used by students around the world. He was also my introduction to the difference between eminence-based and evidence-based science! Opinions or advice that come from a scientist or physician with an established and often stellar reputation but lacking evidence can be termed *eminence-based.* This is in contrast to evidence-based science that is backed by proper studies. In the case of health matters, ideally randomized, double-blind, placebo-controlled trials (RCTs).

Dr. Pauling was a champion of supporting his theories of chemical bonding and molecular structure with evidence drawn from X-ray crystallography. He was a hair away from scooping Watson and Crick on the structure of DNA. And then curiously, in 1970, this eminent scientist who had published hundreds of peer-reviewed articles in the world's leading scientific journals penned a book claiming that the common cold can be cured with large doses of vitamin C. Incredibly, he offered no evidence other than his own and his wife's personal experience. But since Pauling was widely acclaimed as one of the world's leading scientists, the press jumped on the vitamin C story and supplements of the vitamin flew off the shelves. Evidence had been trumped by eminence.

While Professor Pauling's diversion into eminence-based health advice was surprising, the fact is that until the twentieth century, when RCTs began to emerge, medicine, starting with Hippocrates, was essentially all eminence-based. The Father of Medicine owes his fame not to introducing effective cures but to pioneering the idea that diseases have a natural cause and were not doled out by the gods as punishments. However, he wrongly attributed the cause to an imbalance of four body fluids, or humors, namely blood, phlegm, yellow bile, and black bile. Treatments to restore balance included hot and cold baths, purgative herbs, and bloodletting. He did recommend cutting back on food to "starve a fever," but there is no evidence of Hippocrates ever uttering "let thy food be thy medicine," a quote often attributed to him.

Following in Hippocrates's footsteps, Galen of Pergamon furthered the notion that diseases have physical causes and searched for these by carrying out vivisections, experiments on live animals. He made detailed drawings of anatomy, but these contained many errors since dissections of humans were not allowed in ancient Greece or Rome. Galen did gain some insight into human anatomy by examining the wounds of gladiators, but he did not understand the circulatory system, believing that blood was produced in the liver and delivered to tissues where it was dissipated. Galen's eminence was such that his views would hold sway for fifteen centuries until Willian Harvey published his classic work *On the Motion of the Heart and Blood in Animals* in 1628.

The idea that illness is the result of a humoral imbalance persisted until the advent of the germ theory in the 1850s. Until then, eminent physicians like Dr. Benjamin Rush, a signer of the Declaration of Independence, were recommending calomel (mercury oxide) or syrup of ipecac for purging, and bloodletting was a treatment for almost every ailment. George Washington's death was precipitated by his physicians repeatedly bleeding him.

Eminence did not always lead to such brutal methods. Dr. John Harvey Kellogg, for example, achieved fame by claiming to cure people of various illnesses with vegetarian diets, exercise, baths, and yogurt. He was a showman par excellence, famously regaling patients at his "Sanitarium" in Battle Creek, Michigan, with an "experiment" that involved tossing a steak and a banana at a chimp. The ape ignored the steak and happily ate the banana as Kellogg intoned that even this primitive creature knows that meat is not for consumption. Kellogg believed that eating meat was "sexually inflammatory" and maintained that people who ate bacon for breakfast were doomed to masturbate, an activity that would lead to rotting of the brain and insanity. Corn Flakes, according to him, were the anti-aphrodisiacal breakfast food.

In *Rational Hydrotherapy*, a 1,217-page book he published in 1900, Dr. Kellogg claimed that every known ailment can be helped by the application of cold, hot, or tepid water. He described how streams of water directed at various parts of the body were curative and that the soles of the feet are connected by nerves to the bowels, genitals, and brain. There was no evidence for any of this, but Kellogg was an eminent physician, so his claims were not challenged.

Dr. Kellogg's treatments were a cakewalk compared with those of Dr. Walter Freeman, who exemplifies another case of patients placing their faith in eminence instead of evidence. Freeman graduated from medical school at the University of Pennsylvania and later earned a PhD in neuropathology. He came to believe that mental illness is amenable to surgical treatment and invented the transorbital lobotomy, a procedure that involved penetrating the brain with an ice pick-like instrument through the eye socket to sever the connection of the frontal lobes to the hypothalamus. Dr. Freeman had no evidence that the procedure was an effective treatment for mental illness, but incredibly managed to perform some 4,000 lobotomies before he was banned from carrying out the procedure because of the frightening rate of complications. Freeman's eminence was in large part due to having lobotomized Rosemary Kennedy, sister of the future president. Rosemary's erratic mood swings, learning difficulties, and aggressive behavior were deemed unsuitable for a Kennedy by her father, the unsavory Joseph P. Kennedy. He had her lobotomized by Freeman! She ended up unable to walk or speak properly and spent the rest of her life in institutions.

Dr. Luigi Di Bella also carried out lobotomies, albeit metaphorically. He rose to eminence in Italy in 1997 after triggering collective hysteria in cancer patients with his claim that a combination of the hormone somatostatin, vitamins, retinoids, melatonin, and bromocriptine, a Parkinson's disease medication, can cure cancer. Di Bella was a legitimate physician but with an illegitimate claim. His basic ploy was "Trust me, I'm an eminent physician." When oncologists claimed that Di Bella had no evidence, they were accused of conspiring to keep cancer patients from a potentially curative therapy. Demonstrators clamored for "freedom of treatment" and threatened to bring down the government. Officials caved, clinical trials were organized, and the results sent the Di Bella regimen to the junk heap of medicine. Eminence lost out to evidence.

Today we have a whole new breed of physicians whose eminence is due more to their media exposure than their accomplishments. The likes of Drs. Mehmet Oz, Deepak Chopra, Steven Gundry, Joe Mercola, and Christiane Northrup have become mega-influencers and continually use their eminence to upset the evidence-loaded cart.

As for Linus Pauling, he has received a small measure of vindication from a meta-analysis that showed a reduction in the severity of a cold with large doses of vitamin C. When I feel a cold coming on, I have found that taking a gram an hour for four hours can often ward it off. But take that with a grain of salt because I have neither evidence nor eminence.

YOU WANT ME TO STICK A MAGNESIUM ROD WHERE?

"Congratulations! You have in your hands one of the best available tools to increase your health and vitality." So stated the pamphlet that accompanied Dr. Hidemitsu Hayashi's "Hydrogen Rich Water Stick" that I now held in my palm. Several people had asked me if the claims made on behalf of this item were legitimate, so I thought I better order one to see what this purported miracle was all about. It wasn't hard to determine what I was actually holding because the box it came in informed me that the stick was composed of 99.9 percent pure magnesium in an ultra-fine porous polyethylene resin. The reference to "hydrogen rich water" was now clear. It was all about what happens when magnesium reacts with water.

Magnesium is the third most widely used metal in the world, after iron and aluminum. It doesn't occur in nature as the pure metal but has to be produced either from magnesium chloride isolated from sea water or from dolomite, a rock composed of calcium magnesium carbonate, named after eighteenth century French mineralogist Deodat Gratet de Dolomieu who was the first to describe the mineral. The Dolomite mountain range in Northern Italy derives its name from the large amount of dolomite found there.

As any student of chemistry knows, magnesium reacts readily with water to produce hydrogen and magnesium hydroxide. Some of the hydrogen will dissolve in the water, which can then be described as *hydrogenated*. Since the solubility of hydrogen is very low, the maximum concentration that can be achieved is about 1.5 parts per million (ppm) which is 1.5 milligrams per liter. It is from drinking the water in which the magnesium stick has been immersed that the so-called benefits of "increased health and vitality" are to be obtained.

Who is Dr. Hayashi and how much weight should be put on his claims that drinking hydrogenated water provides "superior hydration, antioxidant power, and extra energy?" His statement that "hydrogen is nature's best antioxidant and the only fuel the body recognizes" certainly raises an eyebrow. Hydrogen may be an excellent fuel for rockets and cars, but as far as the body goes, our major source of energy is glucose. One would think that Dr. Hayashi would know that since, as it turns out, he used to be a cardiac surgeon before he pivoted to marketing hydrogenated water. I don't know what it is about cardiac surgeons. Both Dr. Mehmet Oz and Dr. Steven Gundry were respected cardiac surgeons before they became salesmen for questionable products.

So, how questionable is the hype for hydrogenated water? That hype is plentiful, as just a few minutes of internet surfing reveals. It purports to boosts immunity, improve brain function, enhance athletic performance, reduce the risk of cardiovascular disease, promote dental health, alleviate allergies, reduce depression, improve skin structure and, needless to say, cure every disease you have ever heard of. Such a litany of supposed benefits triggers immediate skepticism but not dismissal without investigation. That means surveying the scientific literature on the subject.

I must admit that I was surprised by the number of publications on the topic in proper peer-reviewed scientific journals. Literally hundreds of papers since the first decade of this century when researchers, mostly in Japan and China, began to focus on the antioxidant effects of molecular hydrogen. Oxidative stress was already a hot discussion at the time with the understanding that the oxygen we inhale also produces some "friendly fire" when it takes part in the metabolic reactions that are critical to life. This "fire" is in the form of free radicals, highly reactive species that are deficient in electrons and look to steal them from molecules in their surroundings, namely the proteins, fats, carbohydrates, and nucleic acids that make up our cells. Since electrons are the glue that hold molecules together, damage to these essential cellular components has been linked to disease.

Our body doesn't just stand by and let free radicals run amuck. It produces a number of antioxidants such as glutathione and superoxide dismutase and also enlists the help of numerous antioxidants in the diet that range from polyphenols and vitamin C to carotenoids and selenium. This is why antioxidant supplements have become such hot items. However, these supplements are like carpet bombers, neutralizing all free radicals, which may be why they have not delivered on their promise. Biochemistry is a very complicated business and not all free radicals are threats to health. Some have a physiological role as signaling molecules that mobilize the immune system to attack invaders such as microbes and toxins. Could there be some antioxidant that selectively attacks free radicals, such as the hydroxyl radical, that are believed to have only damaging effects?

Molecular hydrogen is just such an antioxidant. At least that is the rationale claimed by its proponents. There is no question that molecular hydrogen, which is composed of two hydrogen atoms, can break apart with each atom combining with two hydroxyl radicals to produce two molecules of water. There is no real proof of this mechanism, but it is the one proposed to explain the benefits of hydrogenated water. Those benefits seem to have some substance, although they are based on laboratory experiments, animal studies, and a few small clinical trials in humans. None of these are particularly compelling but neither can they be dismissed.

Some examples include a study of patients undergoing radiation therapy for cancer who claimed relief from side effects such as nausea, fatigue, and diarrhea by drinking hydrogenated water. Those side effects are supposedly due to the hydroxyl radicals produced by radiation. Another study demonstrated a reduced rejection of a transplanted heart when hydrogen water was consumed. But that was in rats. In a small study of highly conditioned athletes, there was a slight improvement in performance with hydrogenated water. Another study of twenty people with elevated blood pressure, high blood sugar, excess body fat around the waist, and high cholesterol, a cluster known as *metabolic syndrome*, resulted in a reduction in cholesterol and inflammatory markers.

All interesting stuff, but I'm not quite ready to jump on the bandwagon. I have some persistent concerns. It is hard to imagine that the tiny amounts of hydrogen that could enter the bloodstream from water in which it hardly dissolves in the first place could have effects of practical significance. Would not much of the hydrogen be exhaled from the lungs? Then there is the issue of hydrogen gas production by bacteria in our gut when they metabolize fiber. This is a well-known process, and while some of the gas exits in an undignified fashion, some will enter the bloodstream in amounts that I would guess exceed that available from hydrogenated water.

What I can confirm is that Hayashi's stick does indeed produce hydrogen. Given hydrogen's volatility, I have no idea how much of it remains dissolved in the water. Still in view of the large number of publications, albeit mostly of weak quality, I'm not ready to throw the baby out with the bath water. I've even come across studies that claim that baby should be bathed in hydrogenated water. That would require a good number of Dr. Hayashi's water sticks.

FROM THE JUNGLE TO THE OPERATING ROOM

The history of surgery is often divided into an era described as "before Griffith" and one as "after Griffith" based on Dr. Harold Griffith's introduction of curare in 1942 as a muscle relaxant in surgery. This solved a problem that had plagued surgeons since the discovery of anesthesia a hundred years earlier. The dose of ether, cyclopropane, or chloroform that rendered a patient unconscious had little effect on the autonomic nervous system which meant that when an incision was made, muscles would twitch and even go into spasm, creating difficulties for the surgeon. Use of a higher dose of anesthetic was a work-around, but that had problems of its own. Sometimes the patient wouldn't wake up!

Working as an anesthesiologist at Montreal's Homeopathic Hospital, Dr. Griffith was aware of curare's recent introduction as a treatment for the seizures that were a side effect of the drug Metrazol used to treat depression. Curare controlled seizures by impairing muscular activity, but notably had no effect on the heart. Dr. Lewis Wright, a physician at the Squibb pharmaceutical company in charge of marketing curare as Intocostrin for seizures, suggested to Griffith that the drug could be useful as a muscle relaxant in surgery.

Dr. Griffith administered curare without first doing any animal experiments or seeking permission from an ethics committee. It is a different world today, and a surgeon could not introduce a drug on a whim. However, it should be mentioned that Dr. Griffith was aware that an antidote for curare overdose was available, and he had it on hand should it be needed. Physostigmine, isolated from the Calabar bean, had been found by Austrian physiologist Jacob Pal to counter the effects of curare although its mechanism of action was not known. That was only discovered after English pharmacologist Sir Henry Dale determined that curare produces paralysis by blocking the receptor on nerve cells for acetylcholine, the neurotransmitter needed for muscular activity. It is this activity that physostigmine counters by inactivating acetylcholinesterase, the enzyme that normally breaks down acetylcholine. As a result, the concentration of acetylcholine is increased, and it displaces curare from the receptor. Once the molecular structure of curare's active ingredient, tubocurarine, was determined in 1948, chemists were able to produce a number of analogues that performed better and eventually replaced tubocurarine.

Dr. Griffith was a McGill University graduate in medicine and had spent a year at the Hahnemann Homeopathic College in Philadelphia. Whether he ever practiced homeopathy isn't clear, but certainly his position as an anesthesiologist did not involve the use of nonexistent molecules, the hallmark of homeopathy. The name *Homeopathic Hospital* was also a curiosity since it actually functioned as a regular hospital. Perhaps some homeopathic "remedies" were used, but certainly not in surgical cases.

The path of curare from the jungles of South America to the operating room is a fascinating one. As early as 1516, Europeans learned about the use of poisoned arrows by South American natives from the writings of Pieter Martyr d'Anghera, an Italian who chronicled stories he had heard from travelers to the New World. He described how the Spanish Conquistadores had been attacked with poison arrows and gave a fanciful but fabricated account of the preparation of the poison from plants by women and the determination of its potency by how many of the women were found "half dead" from the toxic vapors.

In 1745, French explorer Charles Marie de la Condamine brought the first sample of curare back to Europe after seeing natives hunt small animals using a blowpipe and poisoned darts. He gave some of the poison to physicians at the University of Leyden in the Netherlands who injected it into a cat and found that it produced paralysis. British naturalist Charles Waterton had also encountered curare on his South American travels, and in 1825, together with surgeon Benjamin Brodie, performed a classic experiment that actually laid the foundation for the use of curare in surgery.

Brodie and Waterton injected a female donkey with curare whereupon it quickly stopped breathing and collapsed as its respiratory muscles became paralyzed. The animal's heart, however, kept beating. At this point, Brodie made an incision in the windpipe and used bellows to pump air into the animal's lungs. He kept this up for two hours, when much to his surprise the donkey raised its head and proceeded to get up, apparently none the worse for wear. The experiment had demonstrated that at a sublethal dose, curare was capable of producing paralysis that lasted until the effect of the drug wore off! Clearly curare had therapeutic potential, but further experiments were hampered by the scarcity of the drug.

That problem wasn't solved until American Richard Gill found a job as a salesman for a rubber company that led to his settling in Ecuador where he learned about curare from the native tribesmen he befriended. Unfortunately, a fall from a horse left him partially paralyzed, suffering from painful bouts of muscle spasms. When he returned to the U.S., his physician Dr. Walter Freeman mentioned that muscle spasms were amenable to treatment with curare. This was the same Dr. Freeman who would become infamous for introducing the "ice pick lobotomy" to treat mental illness, a procedure he performed on President Kennedy's sister Rosemary with frightful consequences.

Motivated to gather a sufficient quantity of curare, Gill returned to Ecuador and put together an expedition to seek out the plants from which curare could be extracted. After five months in the jungle and watching natives prepare the arrow poison, he returned with 12 kilos of crude curare. Curiously there is no historical record of Gill using curare to treat himself, but Nebraskan psychiatrist Abram Bennett heard about Gill's exploits from Freeman and contacted him for a sample. He had in mind to mitigate the side effects of Metrazol that he had been prescribing. When this was successful, the Squibb Company bought all of Gill's curare, and Horace Holaday, one of its chemists, found a way to produce a standardized version that was then marketed as Intocostrin, the drug that Dr. Griffith used. He reported that "within one minute it made the abdomen as soft as dough." The rest, as they say, is history.

THERE IS NOTHING MEMORABLE ABOUT MEMORY SUPPLEMENTS

Imagine what it would be like to lose your memory. Of everything! And even worse, to be unable to form any new ones. That is the nightmare Englishman Clive Wearing has been living since 1985, when the herpes simplex virus — one that normally only causes skin blisters — attacked his hippocampus, the brain structure where memories are stored. Actually, he can't even remember that he is living a nightmare, because the only memory he has is of the past thirty seconds. The exception is that he recognizes his wife, but if she leaves the room and returns a minute later, he greets her as if she had not been there before at all. Even more curious is that Clive, who had been a musician and conductor of note before his affliction, has retained his ability to read and play music.

Wearing's case is extremely unusual, perhaps even unique. However, memory loss to a greater or lesser extent with aging is common, and the hope of prevention has stimulated a great deal of research. It has also stimulated a multi-billion-dollar industry of "brain health supplements," mostly of highly questionable efficacy. There has also been much interest in the possible role of specific foods and diets in preventing age-related memory loss.

As far as supplements go, the weak or nonexistent evidence is dwarfed by the overly exuberant hype. Personal testimonials from people who seem to have only an initial for a second name abound with rosy accounts of regained mental clarity and vague references to clinical trials. Of course, just having carried out a clinical trial says nothing, unless that trial produces a positive result. To be sure, there have been trials galore, but most have used either cultured cells in the lab or rodents navigating mazes. Any human intervention studies have been small, with results of doubtful practical significance. While there are literally hundreds of brain supplements on the market, they all make use of one or more of some twenty ingredients.

Rationale for their use can sound seductive. The most common justifications include increasing levels of the neurotransmitter acetylcholine (huperzine, choline, lecithin, alpha-glycerophosphocholine, *Bacopa monnieri*); decreasing the formation of proteins such as tau and beta-amyloid that interfere with neurotransmission (lion's mane, turmeric, ashwagandha, cinnamon); strengthening myelin, the protective sheath around nerves (phosphatidylserine); improving the fluidity of cellular membranes to improve neurotransmitter function (omega-3 fats); preventing cell death (ginger); nebulous neuroprotective effects (vinpocetine); modulating calcium levels in the brain (apoaequorin); or serving as antioxidants (*Ginkgo biloba*, coenzyme Q10, polyphenols derived from tea, cocoa, or fruits). Many of these ingredients claim to have a combination of these effects.

There is no question that free radicals generated from oxygen during normal metabolism can damage nerve cells through what is termed *oxidative stress* and that antioxidants can mitigate this effect. Neither is there doubt that neurotransmitter activity is critical to cognition, or that beta-amyloid deposits impair memory. But demonstrating that "brain enhancers" can affect these processes in a clinically positive way requires evidence. While they do appear to be safe, such supplements have shown either no or marginal benefit when subjected to randomized, double-blind trials. *Ginkgo biloba*, omega-3 fats, turmeric, and green tea and cocoa extracts are prime examples of failures.

One of the most popular substances, apoaequorin, sold as Prevagen, is advertised as being backed by clinical trials. That "backing" took some imaginative data mining. The company clearly admits that "no statistically significant results were observed over the entire study population," but torturing the data yielded a subgroup of participants who showed a mild benefit. In all probability, this was a statistical quirk since apoaequorin is a protein that is likely to be digested before it has any chance of making it into the brain. There is also hype about apoaequorin occurring naturally in jellyfish. This feeds into the myth that "natural" means safe and effective. Actually, the apoaequorin in

Prevagen is not extracted from jellyfish but is made in the lab, not that this matters. What matters is the lack of evidence of efficacy and the numerous reports of side effects and the launching of federal lawsuits against the company.

There is no shortage of studies that have examined the role of specific foods on brain health. In one randomized study, one group of subjects consumed 60 grams of walnuts, pistachios, cashews, and hazelnuts per day while a control group ate no nuts. The nut group showed increased blood flow to the brain and outperformed the control group by 16 percent on a verbal memory test. In another study, participants who took 13 grams of strawberry powder daily made fewer errors on a word test and reported fewer symptoms of depression. "Eating more kiwi fruit could boost your mental health in just a few days," claims a headline above an article that reports on a study that purported to show subjects who eat two kiwi fruits a day show an improved mood after just four days. Pretty soft stuff.

Now let's turn to some studies with somewhat harder evidence. The Cocoa Supplement and Multivitamin Outcomes Study (COSMOS) found that a group of subjects over age 60 who took a multivitamin-mineral supplement showed a modest benefit in cognition compared with a placebo group. However, other studies, such as one that tracked 6,000 male physicians over twelve years, showed no difference between those who took a multivitamin or a placebo. The Nurses' Health Study that has followed some 48,000 nurses since 1984 found that subjects who had adequate protein intake during middle age, especially plant protein, had better health outcomes, including mental health, as they got older. The researchers suggest nuts for snacks and several meals a week featuring beans, lentils, peas, tofu, and whole grains.

The same Nurses' Health Study also furnished perhaps the most compelling evidence that links diet to a successful battle against cognitive decline. In this case, the focus is on dietary flavonoids, a general family of compounds found in plants with some common features in their molecular structure that account for their antioxidant activity. Intake of flavonoids was calculated from food frequency questionnaires and related to "subjective cognitive decline" as ascertained from questions such as "Do you have more trouble than usual remembering recent events," "Do you have more trouble than usual remembering a short list of items," or "Do you have more trouble than usual following a group conversation or plot in a TV program?"

While such observational studies cannot prove a cause-and-effect relationship, the data do suggest that including several daily servings of foods high in flavonoids such as oranges, peppers, apples, strawberries, and blueberries can slow cognitive decline by as much as 20 percent.

What then is the overall conclusion about slowing memory decline? There is not enough evidence to back any memory supplement. Better spend the money on flavonoid-rich foods. Making sure that daily protein intake is at least 0.8 grams per kilogram of body weight — preferably in large part plant protein — is important.

And how many studies did I tussle with to come to these conclusions? I can't remember. Better stock up on berries.

A NOBEL LAUREATE AND MARATHONERS

"Find me a cure," implored Otto Radnitz in 1920 when his daughter Gerty graduated from medical school in Prague, then part of the Austro-Hungarian Empire. Radnitz had developed diabetes, quite ironic since he was a chemist who managed a sugar refinery and had developed a method for refining sugar. Gerty did not find a cure for diabetes, but according to family lore her father's remark stimulated an interest in the metabolism of sugar that would launch a career culminating in her being the first woman to be awarded the Nobel Prize in Physiology or Medicine. The honor, "for their discovery of the catalytic conversion of glycogen," was especially memorable because it was shared with her husband, Carl Cori, one of the rare times that the prize was awarded to a couple.

Gerty and Carl met when they entered medical school, but their relationship was put on hold when Carl was drafted into the Austrian army during the First World War. He would later describe how that experience motivated him to research disease as he witnessed the influenza epidemic sweep through the troops with doctors unable to offer any help. After the war, Carl returned to university, and the couple were married and moved to Vienna where Carl found a research position and Gerty worked in a hospital's pediatric ward. Life in Vienna was difficult and with the rise of antisemitism and Gerty's Jewish heritage, the couple looked for greener pastures and found it in Buffalo, New York, where Carl was hired as a biochemist at the State Institute for Malignant Disease and Gerty as an assistant pathologist.

The director of the institute frowned on the couple working together and even threatened to fire Gerty if the collaboration continued. However, once the couple started to publish pioneering papers about carbohydrate metabolism, the opposition faded. Offers from universities began to pour in, but only for Carl. One even opined that it was "un-American" for couples to work together. Washington University in St. Louis was smart enough to ignore the institution's nepotism policies and hired Gerty as well, but with one-tenth the salary her husband was to receive. It would be thirteen years before her contributions were recognized and she was finally promoted to a position of full professor with the appropriate salary. The Coris would go on to mentor six eventual Nobel Prize winners, an incredible achievement.

So impactful was their work that in 2004 the American Chemical Society would recognize the Coris' lab as a National Historic Landmark designating it as "the place where pioneering research was carried out that led to the understanding of the metabolism of sugars and elucidated the 'Cori cycle,' the process by which the body reversibly converts glucose and glycogen, the polymeric storage form of this sugar." That sounds complicated, so let's try to simplify why this research received numerous accolades, including the ultimate one, the Nobel Prize.

The human body is one of the most complicated machines on the face of the earth. Unraveling the details of how it works is perhaps the greatest challenge faced by scientists. It is here that the Coris made their contribution by explaining how glycogen, a molecule that is a chain of glucose units, serves as a source of energy. Everything we do, whether it is typing on a keyboard or running a marathon, requires energy. That energy is provided by a process that in a simplified way can be described as combustion. Any combustion process requires a fuel that reacts with oxygen to yield water and carbon dioxide with the concomitant release of energy. The body relies on the conversion of glucose and fats into a molecule of adenosine triphosphate, or ATP, that actually provides the energy required. But interestingly, the "burning" of fats involves a step that requires glucose, which is why the expression "fats burn in the flame of carbohydrates" is familiar to endurance athletes.

The critical glucose is supplied by carbohydrates in the diet, but in order to keep blood levels constant, excess glucose molecules are linked together to form glycogen, a polymer that is then stored in muscles and in the liver to be called upon to supply glucose should blood levels drop. Marathon runners, for example, know that they have to maximize their glycogen storage by loading up on carbohydrates before the race, and that they have to consume glucose during the race in the form of gels or drinks to ensure they have enough fuel. Everything is fine as long as they keep a pace that allows enough oxygen to be inhaled to support the combustion process. However, should they at some point end up running at a pace that prevents sufficient oxygen uptake, their body can no longer burn glucose or fats in the normal way and has to switch to a different mechanism whereby energy can be derived from glucose without the use of oxygen. This is referred to as *anaerobic metabolism.*

In this process, glucose generates ATP by being converted into lactic acid. But the glucose in the blood is quickly used up and has to be replenished by the breakdown of muscle glycogen. Soon, the stores of this run out as well. But there is a backup process, as the Coris discovered. The lactic acid that builds up in the blood can be converted back to glycogen in the liver, from where it can resupply the blood with glucose. But this cannot happen indefinitely and eventually blood glucose drops to a level when the athlete "hits the wall." Sudden fatigue appears along with a sensation of the legs turning to jelly. Disorientation can ensue as the brain is robbed of its supply of glucose.

These are the complex processes that were deciphered by the Coris. Their efforts are even commemorated in a U.S. stamp that features the molecular structure of glucose-1-phosphate, the critical intermediate in the conversion of glucose into glycogen according to the Cori cycle. Amazingly, the molecular structure, also known as the *Cori ester*, depicted on the stamp is wrong! The Coris were spared this annoyance since the stamp was issued in 2008, long after they had both passed away. Although Carl lived to eighty-eight, Gerty was diagnosed with a fatal disease of the bone marrow the same year she was awarded the Nobel Prize, and she died at sixty-one.

Any marathoner or cyclist who has made use of the timely consumption of carbohydrates to avoid hitting the wall, or "bonking" as it is also known, owes gratitude to the Coris, a remarkable husband and wife team.

MAIMONIDES AND JEWISH PENICILLIN

"As the Good Book says, when a poor man eats a chicken, one of them is sick." That line comes from Tevye in the classic musical *Fiddler on the Roof*. The village Rabbi's son overhears this and asks: "Where does the Book say that?" to which Tevye responds, "All right, all right! It doesn't exactly say that, but someplace, it has something about a chicken." And he is right! The "Good Book" to which Tevye refers is the Talmud, a collection of writings compiled in the fifth century by Jewish theologists about ethics, philosophy, religious observance, rituals, dietary laws, and traditions that serve as a guide for the conduct of daily Jewish life. One of the discussions in the Talmud mentions "the chicken of Rabbi Abba which for medical reasons was cooked so thoroughly that it completely dissolved." Sounds like the sage considered chicken soup as medicine!

Actually, credit for the first mention of chicken soup as "medicine" goes to a Chinese document dating to the second century BC in which the soup is described as a "yang food" that warms the body and has an invigorating effect. But it was Moses Maimonides, the twelfth century Jewish philosopher and physician, who brought the healing properties of chicken soup into the limelight. Maimonides was born in the Iberian town of Cordoba, but his family had to flee when it was conquered by the Almohades, an extremist sect that forced Christians and Jews to convert to Islam or face death. Eventually the family ended up in Egypt where Moses became a highly respected physician, even serving in the Sultan's court. Exactly how he was educated is not clear, but we do know that his medical writings reflect Galen and Hippocrates's view that health is a function of the balance of the body's four humors, namely blood, phlegm, yellow bile, and black bile.

Although the humoral theory had no scientific basis, Maimonides's recommendations to bring them into balance had value. Well ahead of his time, he emphasized the importance of clean air, clean water, a healthy diet, and exercise and stated that "a physician should begin with simple treatment, trying to cure by hygiene and diet before he administers drugs." That diet included the meat and broth of chickens that he claimed "Rectified corrupted humors especially black bile that causes melancholy." He even recommended chicken soup as a medication for leprosy. That would not have had any effect on the bacteria that cause the disease but could have provided nourishment for convalescence. Maimonides was particularly fond of chicken testicles for convalescence and also claimed that "they aid the libido in a strongly perceptible manner." While there is no record of Maimonides specifically addressing chicken soup for the common cold, Dr. Fred

Rosner, the world's foremost expert on Maimonides, has dug up a quote that "soup made from an old chicken is of benefit against chronic fevers that develop from white bile and also aids the cough which is called asthma."

Whether due to the Talmud or to Maimonides's conjectures, by the early 1900s in Russia, the era of *Fiddler on the Roof*, the mythology of chicken soup as medicine was well established and the soup had become a staple at the Shabbat dinner. But is *mythology* really the right term? Could there be some science to the soup's supposed benefits? That question has tickled the fancy of a number of researchers, some perhaps motivated by the chance to garner headlines, something that any story about chicken soup is guaranteed to do.

First out of the block were physicians at Mount Sinai Medical Center in Miami who in 1978 decided to investigate whether "chicken soup, a treatment long advocated by Jewish mothers," was effective for alleviating upper respiratory tract ailments. In fifteen healthy patients they devised an ingenious way using tubes, tiny Teflon discs inserted into the nose, and scuba diving masks to measure the speed at which mucus flowed out of the nose and air flowed in. The subjects were asked to consume hot water, hot chicken soup, or cold water either by sipping or drinking through a straw. Sipping hot water or hot chicken soup both increased nasal mucus velocity as did chicken soup by straw, but hot water by straw did not. None of the treatments changed nasal airflow. Although this study received a great deal of publicity with articles highlighting the increased flow as a result of consuming chicken soup, the fact that hot water had the same effect was hardly mentioned.

The chicken soup literature was silent until 2000 when a study with the title "Chicken Soup Inhibits Neutrophil Chemotaxis In Vitro" once again captured the media's attention. Chemotaxis is the ability of cells to move in a particular direction in response to a stimulus such as chemicals emitted by a bacterium or virus. Neutrophils are white blood cells that are attracted to the site of infection by signals released from cells that have been infected and damaged by bacteria or viruses. The neutrophils then engulf the invading microbe and destroy it. It is the elimination of the breakdown products that results in the runny nose, sneezing, and congestion, the classic symptoms of a cold. *In vitro* literally means *in glass* and refers to experiments done in the lab, as opposed to using animals or people.

The researchers found that chicken soup significantly inhibits neutrophil migration and does so in a concentration dependent manner. Both the broth and the vegetables it contained showed this effect when tested separately. What does this mean? Not much. First, a study in glassware with neutrophils immersed in chicken soup cannot be extrapolated to what may happen in a cold sufferer. It is hard to even guess if whatever "active ingredient" there may be in the soup makes it from the stomach to the respiratory tract. And while slowing neutrophil activity may lessen symptoms, it also lengthens the time to destroy the invading microbe. So really, very scant evidence here for chicken soup as medicine.

If chicken soup isn't that great for the body, can it still do something for the soul? Researchers at the University of Buffalo think so. They hypothesized that the real value of chicken soup is in "acting as a comfort food because of repeated exposure in the presence of relational partners." They showed that subjects who were lonely because of a fight with a friend or breakup with a romantic partner found solace in chicken soup. The soup acted as "social surrogate" for the missing company.

All of this leaves us with the sense that the science of chicken soup is really soft, and that its label as "Jewish penicillin" is more whimsy than fact. But the evidence that it tastes great is pretty hard, and I think we can take comfort in that. The chicken, I suspect, would have a different opinion. There is yet another reason to enjoy a chicken soup in which carrots, parsnips, onions, celery root, and garlic have frolicked. As the Good Book says, "It's tradition!"

KING CHARLES AND THE HOMEOPATH

The British press is gushing with articles about King Charles's appointment of Dr. Michael Dixon as head of the royal medical household. Most accounts are highly critical because of Dr. Dixon's record of promoting various "complementary" therapies such as acupuncture, herbal remedies, and homeopathy. It is his championing of the latter that has raised the most eyebrows since in the eyes of the vast majority of physicians and scientists, homeopathy is scientifically bankrupt. Before going any further, let's clarify that homeopathy, contrary to what most people believe, is not an umbrella term for all treatments that fall into the "alternative" category. It is a very specific practice invented over two hundred years ago by German physician Samuel Hahnemann who was troubled by the brutality of the medical interventions used by physicians at the time in an attempt to drive ailments out of the body. There had to be a gentler way than bloodletting, purging, and the use of toxic substances such as calomel (mercuric chloride) to treat illness, Hahnemann thought.

One remedy at the time that really did work was the use of the ground bark of the cinchona tree imported from South America for the treatment of malaria. The bark contains quinine, a compound that is deadly to the parasite that causes malaria when it finds its way into the bloodstream through the bite of a mosquito. However, the amount of bark to use in a patient was guesswork. Hahnemann, by all accounts a caring physician, tackled this problem by experimenting on himself. He wanted to find out how much cinchona bark could be given to a patient before causing harm and was surprised when he developed a fever typical of malaria after taking a large dose. At that moment the central concept of homeopathy was born. A substance that causes symptoms in a healthy person will cure like symptoms in a sick person. But Hahnemann knew that he had been giving much smaller doses of cinchona bark to his malaria patients than what he had taken, and that resulted in the second tenet of homeopathy. The smaller the dose, the greater the potency. This was a logical non sequitur, but when Hahnemann began to get positive responses from patients whose symptoms he treated with tiny doses of substances that at a high dose he had "proven" caused those symptoms in healthy subjects, he was sold. The remedies seemed to work even better when the solution was shaken between further dilutions, and that became the third tenet of homeopathy.

Today, scientists bristle at the idea of nonexistent molecules having a therapeutic effect. And that is exactly what we are dealing with because with our current knowledge of chemistry it is possible to determine that after a sequence of hundred-fold dilutions repeated twelve times, there is not a single molecule of the original substance left. Because homeopaths now have to admit this, they have forged an alternate explanation for how homeopathy works. The shaking between dilutions leaves an imprint, a ghost if you will, of the original substance in the solution. Not only is there no evidence that the structure of water can somehow be altered in this fashion, there is no explanation offered for how this ghostly image can cure disease. To put it succinctly, homeopathy is scientifically implausible. Its precepts defy the laws of chemistry, physics, and biology. It cannot possibly work.

But it does, claim its proponents! King Charles, Queen Elizabeth, and numerous other royals have sung its praises. So have millions of people in India, Germany, and France. How can this be? Scientists maintain that any benefit of homeopathy can be explained by the placebo effect. No, say homeopaths, as they point to placebo-controlled trials that "prove" homeopathic preparations work. No, they don't prove any such thing. If you carry out enough studies, you will occasionally get positive results by chance alone. That is why in science we don't set store by single studies but look at the totality of evidence. And when we do that, it is clear that homeopathy works no better than a placebo. That, though, does not mean that homeopathy should be swept under the carpet; the placebo response is in the 30 to 40 percent range, certainly not insignificant. But there is a potent caveat here.

Relying on nonexistent molecules is fine when dealing with minor aches and pains, but placebos can only change the perception of a disease, not its underlying cause. If the symptoms are due to a serious ailment, use of a homeopathic "remedy" can delay potentially effective treatments.

Dr. Dixon and King Charles would agree that homeopathy should not be used instead of conventional medicine but as a "complement" to it. They argue for "integrated medicine," which Charles describes as "the best of both worlds." But the issue here is that one of those worlds, namely conventional medicine, strives for evidence, while the world of alternative medicine is satisfied with anecdotes. In truth, conventional medicine is the real integrated medicine. When some treatment is shown to be effective through proper trials it is embraced and incorporated into practice. But when Dr. Dixon or King Charles speaks of the need for integrated medicine, what they mean is that doctors should consider recommending modalities such as reflexology, herbalism, traditional Chinese medicine, Ayurvedic medicine, and homeopathy, all of which lack compelling evidence.

The argument for integrating these alternative approaches often introduces the claim that scientific medicine treats only diseases not patients and is not "holistic." Anyone familiar with current medical education knows that claim to be bogus. The patriarchal days of "just do as I say, doctor knows best" are long gone, and medical students are taught to discuss every facet of a patient's life before coming to a mutually agreed-upon treatment protocol. True, overburdened physicians cannot spend as much time with patients as can alternative practitioners, but the answer to that problem does not lie in asking doctors to legitimize treatments that lack evidence.

Dr. Dixon claims that his professional life "turned from gray to color" when "frustrated by the blunt instruments of his medical training" he gravitated towards offering his patients the likes of acupressure, a range of herbs, meditation, dietary advice, and homeopathic pills. "I have witnessed the beneficial effects in so many patients and that has been proof enough for them and for me." But that is not how science works.

As we are fond of saying, the plural of anecdote is not data. As far as data goes, we have it for the benefits of meditation, diets, and even some herbs, and these are by no means solely in the domain of alternative practitioners. But homeopaths promoting the idea that something that contains nothing can cure something misleads patients. Homeopathy is pure folly. If you have been prescribed a homeopathic remedy, you better not forget to take it because if you do forget, you will overdose!

A TOMATO MYTH

Salem, New Jersey, a Saturday in August. Dressed in period costume, a man strides up the stairs in front of the courthouse, faces the crowd that has gathered, reaches into a basket, picks up a tomato, and proceeds to take a bite. The onlookers cheer and burst into applause. They are watching a re-enactment of a historic event that took place in 1820 when Robert Gibbon Johnson ate a tomato in full view of an audience that expected him to foam at the mouth, twitch, and then expire. That didn't happen! Johnson had proven that the "poison apple," as the tomato was known at the time, was not poisonous after all. And so, as the story goes, the tomato-growing industry in America was born!

A great story, repeated in numerous articles and books. But it presents a problem. There is no evidence that it happened. Johnson was indeed a real person, a horticulturist and founder of the Salem Historical Society, but there is no record of him having any special connection to tomatoes. It seems that Salem postmaster and amateur historian Joseph Sickler cooked up the captivating account a hundred years after the supposed event in order to bring attention to Salem. And that it did! The story appeared in Stewart Holbrook's 1946 book, *Lost Men of American History*, and really took wings in 1949 when the CBS radio show *You Are There* broadcast the re-enactment live.

The cloud of being poisonous really did hang over the tomato in the early nineteenth century. There are two accounts of the origin of the toxic tomato tale, an apocryphal one involving European aristocrats, and a more realistic one of mistaken identity.

Tomatoes were introduced into Europe by the Spanish Conquistadors, but due to their scarcity could only be afforded by the wealthy who commonly ate from dishes made of pewter, an alloy of tin and lead. As the story goes, the acids in the tomato leached lead from the pewter and resulted in the demise of the diner. While tomatoes are indeed acidic, the probability of the trace amounts of lead leached from pewter causing death is on par with the probability of giant tomatoes attacking humans as in the 1978 cult film *Attack of the Killer Tomatoes.*

Today, Italians are huge consumers of tomatoes, so it is ironic that the poison legend traces back to sixteenth century Italian herbalist Pietro Andrea Matthioli who classified tomato as part of the deadly nightshade family. The tomato plant does resemble the belladonna bush, the berries of which were known to be poisonous. While the leaves and stem of the tomato plant do contain some of the alkaloids that make belladonna berries toxic, its fruit does not. And indeed, the tomato is a fruit, according to the definition of a fruit being the sweet and fleshy product of a tree or plant that contains the seeds needed for reproduction.

Amazingly, it was a ruling by the U.S. Supreme Court in 1893 that declared the tomato to be a vegetable, not a fruit. A prevailing law at the time required a tax to be paid on imported vegetables, but not fruit. The John Nix & Company was a large importer of tomatoes and argued that since the tomato was botanically a fruit, it should be exempt from the tax. However, the court decided that in this case applying the dictionary definition of a fruit is not appropriate and the tomato should be classified according to how it is perceived by the public which is as a vegetable. Today it is New Jersey's official state vegetable.

Real credit for popularizing the tomato goes not to Robert Gibbon Johnson, but to Dr. John Cook Bennett, a quixotic physician with an unsavory history of having launched a medical diploma mill, selling medical degrees to anyone who could come up with ten dollars. He also ensnared women in sexual relations, declaring that he had been sanctioned by God to practice "spiritual wifery." Bennet claimed to have traveled through Europe and seen that tomatoes cured diarrhea and indigestion and prevented cholera. He recommended that tomatoes replace calomel because they were less harmful, predicting that "a chemical extract will probably soon be obtained from it which will altogether supersede the use of calomel in the cure of diseases." Although his claims of the medicinal value of tomatoes were illusionary, replacing calomel, a concoction made with toxic mercury chloride, was a good idea.

Bennett's prediction of a tomato extract came to fruition in 1837 when Dr. A.J. Holcombe of Alabama introduced his tomato pills "possessing hepatic, cathartic, and diuretic qualities." Holcombe's pills were soon joined by those produced by "Dr." Archibald Miles and Dr. Guy Phelps triggering a lusty advertising battle between the two men. Miles, who actually had no medical education, called Phelps a "quack" and a "charlatan," prompting Phelps, a Yale-trained physician, to retort that Miles had "about as much claim to the title of doctor as my horse." The battle, with the combatants hurling insults of "fraud" and "copycat" at each other, petered out after a couple of years when allegations emerged that neither pill actually contained tomatoes. Still, for a period the pills did supply people with a dose of placebo that was certainly preferable to calomel.

The "tomato pill" reemerged in a different guise in the twentieth century when epidemiological studies indicated a reduced risk of prostate cancer in men with higher consumption of tomatoes and tomato products. The theory was that lycopene, the compound responsible for the fruit's red color, had anticancer properties, and lycopene supplements appeared on shelves. While there appears to be substance to the tomato's role in reducing the risk of prostate cancer, studies using lycopene supplements have disappointed. It seems that it is the collage of the numerous compounds found in tomatoes that produce the benefits. Those benefits may even extend to the cardiovascular system, as demonstrated by a recent study of the dietary habits of over seven thousand seniors. Consumption of just one large tomato a day was associated with a significant decrease in the risk of hypertension. Although there is the usual caveat that tomato consumption may just be a marker for a healthy diet that includes lots of fruits and vegetables, there is enough noise in the scientific literature about the benefits of tomato consumption to pay attention.

Maybe we should all be re-enacting Robert Gibbon Johnson's act of bravery . . . every day.

A TARIFF ON PEANUTS YOU SAY?

Slap a tariff on peanuts! That was the decision arrived at by the U.S. Congress in 1921 after hearing testimony about how American peanut farmers were being undercut by imported peanuts from China. The witness was George Washington Carver who also expounded on all the uses to which peanuts could be put. It was unusual at the time for an African American to appear in front of a congressional committee, but Carver, born around 1864 and raised by Missouri farmers who had owned his mother as a slave, had made a name for himself as an agricultural expert. After graduating from the University of Iowa he had taken up a position as head of the Department of Agriculture at the Tuskegee Institute in Alabama. To underline the importance and versatility of peanuts, Carver had brought along samples of ice cream, candy, instant coffee, milk, oil, ink, and a face cream, all made from peanuts. Indeed, in his career he would go on to formulate more than 300 items from peanuts, including soup, doughnuts, shaving cream, laxatives, and laundry soap.

"Here is a breakfast food," Carver pointed to the table in front of him. "I am very sorry that you cannot taste this, so I will taste it for you" he quipped, eliciting much laughter. The committee members were so impressed with Carver's presentation that his allotted time of ten minutes was repeatedly extended. The "Peanut Man," as he would eventually be known, concluded by saying that he was unaware of a single case of anyone being hurt by peanuts. That is not surprising given that mentions of food allergy did not appear in the medical literature until the 1920s. Only in the 1990s would peanut allergy be recognized as a serious problem.

At Tuskegee, Carver had carried out research on crop rotation that changed the face of Southern agriculture. Cotton, the dominant crop in the South, presented difficulties because it depletes soil nutrients and requires a lot of fertilizer, which most farmers could not afford. But Carver found that rotating cotton with legumes solved this problem. Legumes harbor microbes in their root nodules which can fix nitrogen, meaning that they can convert nitrogen in the air into nitrogen compounds that remain in the soil and serve as nutrients for subsequent plantations. Soybeans and peanuts were ideal for this soil enrichment.

George Washington Carver's name is intimately and justifiably linked with peanuts, but contrary to many accounts, he did not invent peanut butter. Centuries earlier, the Incas were already grinding peanuts into a paste, and in 1884, Montreal chemist Marcellus Gilmore Edson filed a patent for the manufacture of a peanut paste that had "a consistency like that of butter, lard, or ointment." It was to be combined with sugar to make a candy, but there is no evidence the paste was ever sold as a "butter." Another candidate for the invention of peanut butter is St. Louis physician Dr. Ambrose Straub, who, sometime in the early 1880s, crushed peanuts into a high protein paste for his patients who could not chew because they were missing teeth. Straub's friend George Bayle owned a food processing plant and produced the "nut butter" that was then offered to the public at the Chicago World's Fair in 1893. Dr. John Harvey Kellogg, of cereal fame, also claimed to be the inventor of peanut butter based on having filed a patent in 1896 for boiling nuts and passing them through rollers to produce a paste. Kellogg promoted a vegetarian diet in which his nut butter replaced meat, which he claimed encouraged sexual activity that robs the body of energy.

These early versions of peanut butter did not keep well because the liquid oil separated from the solid carbohydrates and proteins. This problem was solved by Kentucky entrepreneur Joseph Rosefield who in 1922 patented a partial hydrogenation process to solidify peanut oil and keep it from separating. The process by which hydrogen gas in the presence of a nickel catalyst converts the carbon-carbon double bonds in liquid unsaturated fats into the single bonds of solid saturated fats had been introduced in 1902 by German chemist Wilhelm Normann. Rosefield's partial hydrogenation resulted in the desired texture for peanut butter, but unknown at the time, it also introduced a side product, the "trans fats" that would become notorious for increasing the risk of heart disease. Today, partially hydrogenated fats are no longer used in foods, having been replaced by fully hydrogenated vegetable oils or palm oil. These fats, as well as the sugar that is often added to peanut butter, also have issues, but that is not a concern when only a couple of spoonfuls of peanut butter are incorporated into the daily diet. "Natural" peanut butters that contain only peanuts are also available. With these, the peanut oil may rise to the top, but that can be remedied with a bit of vigorous stirring.

While there is no significant nutritional issue with peanut butter, allergy to peanuts can affect up to 3 percent of children in Westernized countries. Peanut allergy, although not as prevalent as allergies to eggs or milk, is outgrown in fewer than 20 percent of cases and can have deadly consequences. The most severe reaction is anaphylaxis, which can lead to life-threatening respiratory failure unless reversed with an intramuscular injection of adrenaline (epinephrine). Anyone with a peanut allergy must have an EpiPen within reach at all times in case of accidental exposure.

Why food allergies, especially to peanuts, have risen significantly in recent decades is unknown. One clue may come from the discovery that sensitization may not require ingestion but may occur from exposure through the skin. This can happen if peanut products in a house inadvertently contact a baby's skin. It has been suggested that washing babies every day, a relatively recent practice, may change the skin's permeability to foreign proteins.

Other possibilities that have been raised to explain the increase in food allergies include changes in the intestinal microbiome during infancy and decreased exposure to infectious agents during childhood that makes the immune system less "busy." With fewer infectious agents to fight, the immune system may unleash its weaponry against targets that do not actually pose a risk. These are only theories, but it has become clear that delaying exposure to peanuts does not reduce the risk of allergy.

Studies have shown that in infants with severe eczema or egg allergy, conditions that increase the risk of peanut allergy, risk can be reduced by introducing peanut-containing foods into the diet as early as four to six months of age. If there is no eczema or food allergy, guidelines recommend that peanut-containing foods can be introduced along with other solid foods.

I have no personal experience to recount here because peanut butter was unknown in Hungary when I was growing up. However, its popularity, inexplicable to my palate, did prompt me to look into the history of this American favorite that introduced me to the exploits of George Washington Carver. I now understand his epitaph, which reads: "He could have added fortune to fame, but caring for neither, he found happiness and honor in being helpful to the world."

WHATEVER HAPPENED TO LORENZO?

Given that he had trained and practiced as a physician before turning to filmmaking, it is no surprise that George Miller was so captivated by the story of Michaela and Augusto Odone's struggle to save their son from a deadly disease that he decided to turn it into a movie. *Lorenzo's Oil*, starring Nick Nolte and Susan Sarandon, became one of 1992's most successful films and introduced the public to the many nuances of medical research and the contribution that can be made by the sheer doggedness of parents who dedicate their lives to finding a cure for their son stricken with adrenoleukodystrophy (ALD), a devastating disease.

In short, the film is about a young boy afflicted with a horrendous, progressive, and incurable disease whose parents immerse themselves in the study of biochemistry and discover an effective treatment, which is not recognized by an overcautious and callous medical establishment. Lorenzo was an exceptionally bright child until in 1984, at the age of six, he began to exhibit changes in his behavior. His movements became awkward, and he began to have trouble hearing. Then came slurred speech, temper tantrums, and difficulty keeping his balance. The diagnosis was terrifying. Lorenzo was a victim of ALD, a genetic disease that typically shows up in boys between the ages of four and ten and leads to death within a few years.

ALD is the result of very long chain saturated fatty acids, containing more than twenty-two carbons, that build up in blood plasma. These fatty acids are normally produced from dietary shorter chain fatty acids by enzymes appropriately called *elongases* and are then transported into cells by special transporter proteins. Within cellular organelles known as peroxisomes, the long chain fatty acids are broken down into smaller molecules that perform a variety of vital functions. However, in ALD, a mutation in a gene prevents the formation of the transporter protein and results in the buildup of the long chain fatty acids in blood that end up inactivating nerve cells by stripping them of myelin, their protective sheath.

The Odones did not accept the prevailing medical opinion that no treatment for ALD exists and refused to just wait and watch Lorenzo die. They began to haunt libraries and scoured the medical literature for papers about the disease until one day they came upon a study in a relatively obscure Polish journal describing how the concentration of long chain fatty acids in the bloodstream of rats can be lowered by feeding the animals a diet of other fatty acids. It is at this point that the Odones convinced Dr. Hugo Moser (portrayed as Dr. Gus Nikolais by Peter Ustinov in the film), the world's leading expert on ALD, to convene the first ever conference on the disease. Here they learned from Dr. William Rizzo of the Medical College of Virginia about preliminary experiments using olive oil to lower plasma levels of the troublesome long chain fatty acids. It seems that the elongase enzymes that normally produce these from shorter saturated fatty acids can also interact with unsaturated fatty acids. If they are busy metabolizing these, then the long chain fatty acids are not produced.

The Odones were quick to start feeding olive oil to Lorenzo, and indeed, the blood levels of his long chain fatty acids dropped, although not dramatically. They then began a search for oils that might have an even greater affinity for the elongase enzymes and discovered that erucic acid from rapeseed was a candidate. The oil was very expensive to produce, but a British lab agreed to supply it to Lorenzo for experimental purposes. Feeding Lorenzo a combination of erucic acid and oleic acid from olive oil resulted in a significant reduction in his blood levels of long chain fatty acids. Unfortunately, by this time Lorenzo was totally paralyzed, could not speak, hear, or see, and was only able to communicate somewhat with eye movements. However, to the surprise of his doctors, not only did Lorenzo manage to evade his predicted demise, but by the age of fourteen actually showed some improvement. He was able to swallow, regained some sight, and managed to move his head, allowing him to communicate by means of a computer.

The medical community was not thrilled with the movie, claiming that the Odones' discovery of the oil was overly romanticized and Lorenzo's "improvement" overstated. Furthermore, doctors were portrayed as arrogant, close-minded, and cold-hearted, unwilling to help in what they considered to be a futile search for a cure. Augusto Odone underlined this view in the film with his comment that "these scientists have their own agenda, and it is different from ours." At one point, the character of Dr. Moser declares that "I will have nothing to do with this oil." That was an unfair attribution as pointed out by Dr. Moser in his review of the film published in the British medical journal *The Lancet*. The film, he maintained, invented conflicts between the Odones and the medical establishment that did not exist.

Far from dismissing the nutritional treatment of ALD, Dr. Moser went on to conduct trials with Lorenzo's Oil on a large number of ALD patients. The results were disappointing. It seems that Lorenzo with his unusual longevity was an outlier, his survival perhaps due more to the extraordinary care and attention he received from his parents than to the oil administered through his feeding tube. Although Lorenzo's Oil has not turned out to be a cure for ALD, Dr. Moser did find a silver lining in the cloudy story. In boys who have the genetic variant that can lead to ALD, the disease can be prevented if treatment with Lorenzo's Oil is started before symptoms appear.

Although the film's portrayal of the Odones as lay people who discover a therapy that had eluded the scientific community plays somewhat loose with the facts, profits from the film did allow the Odones to set up the Myelin Project, a foundation to promote research into demyelinating diseases. All the personalities involved in the Lorenzo's Oil story have since passed on, but Lorenzo did live to the unprecedented age of thirty. Just how much of a life that was is an open question.

THE PSEUDOSCIENCE OF PHRENOLOGY

Walk through a pharmacy today and you may encounter a device that automatically reads blood pressure. A hundred years ago that encounter may well have been with a psychograph that analyzed your personality traits and predicted success or failure in life. After donning this metal helmet equipped with numerous rods that pivoted to match the shape of your skull, you would walk away with a printout that assessed your mental attributes on a scale of one to five and offered advice on making changes. A typical reading of three for "wit" was: "Try to get fun and mirth out of life. Smile and joke with others to improve your wit. You need to appreciate more of the ludicrous in life." Hmmm. I think anyone putting on a psychograph was already into the ludicrous.

This bizarre contraption was invented in 1905 by Henry Lavery, capitalizing on the public's interest in phrenology, the brainchild of German physician Franz Joseph Gall. Introduced in the late 1700s, phrenology was based on the belief that the shape of the skull reflects a person's mental abilities and reveals specific character traits. Gall came to this conclusion in his school days upon noting that the most intelligent student in his class had noticeably prominent eyes and a large forehead. Gall rationalized that just like muscles that increase in size with exercise, parts of the brain that are most used expand and distort the skull, producing bumps that are then amenable to analysis. Intelligence, mathematical skills, musicality, and numerous characteristics such as love of children and desire to own material things could be determined by studying an individual's skull.

Phrenology became all the rage and even writers such as Walt Whitman, Emily Bronte, Edgar Allan Poe, and Sir Arthur Conan Doyle bought into the craze. The creator of Sherlock Holmes embracing phrenology was particularly curious, since the detective, the very embodiment of logic, was a tireless promoter of basing theories on facts. Conan Doyle depicted both Sherlock Holmes, as well as his highly intelligent arch-enemy Professor Moriarty, as having dominant foreheads. Upon finally encountering Holmes, Moriarty remarks that Holmes has less frontal development than he would have expected. Interestingly, Conan Doyle himself had a prominent forehead. The author, who hailed from Edinburgh, could well have been influenced by George and Andrew Combe, who had founded the Edinburgh Phrenological Society. The building still stands today, decorated with the busts of the Combes as well as those of Gall and his disciple Johann Spurzheim who coined the term *phrenology*.

Spurzheim claimed that diagnosis of traits by examination of the shape of the cranium was important because these traits could be altered. People whose skulls showed signs of weak intelligence were not doomed, he maintained. If they were identified, and provided with a means of education, the appropriate part of their brain would be exercised and would increase in size, boosting their intelligence.

Spurzheim's lectures popularized phrenology first in Victorian England then in America where his methods were adopted by the brothers Lorenzo and Orson Fowler who became noted phrenologists and even commercialized Fowler heads, busts marked with areas representing different attributes. These were widely used by phrenologists to interpret their findings as they palpated their subjects' skulls with their fingers. Companies based hiring practices on phrenological analysis, marriage partners were selected based on skull shapes, and parents were keen to learn about their children's future prospects by putting them in the hands of a phrenologist. Lorenzo later hung out a shingle in London that led to an encounter with Mark Twain who poked fun at Fowler by remarking that the phrenologist had found on his skull a cavity where humor ought to be. Many cartoonists also had fun at the phrenologists' expense, and to most scientists the suggested link between bumps on the head and innate characteristics seemed absurd and worthy of being mocked.

Science, however, is not based on "seems." It is based on investigation and the unearthing of facts. Coming to a conclusion must be based on evidence, not on a judgment of plausibility. Somewhat surprisingly, phrenology was not put to a scientific test until 2018, when researchers at the University of Oxford used magnetic resonance images (MRI) accumulated by the U.K. Biobank's imaging study to search for any possible relationship between the shape of the scalp and personality traits. They were able to do this because the Biobank study also includes data from questionnaires that assess cognitive abilities as well as lifestyle measures such as a history of the number of sexual partners. A second part of the study involved investigating whether the contours of the skull were a reflection of the characteristic folds of the brain.

One of Franz Gall's claims was that the "Organ of Amativeness," described as "the faculty that gives rise to the sexual feeling," was located in the brain near the nape of the neck and could be located by a protuberance on the skull. The size and shape of the bump was said to be a measure of sexual desire. How did he come to this conclusion? By probing the skulls of some "emotional" young women, two recently widowed neighbors, and his local vicar, subjects who supposedly represented the spectrum of sexual desires. The Oxford researchers found no correlation between the bumps on the skull that supposedly locate the "Organ of Amativeness" and the number of sexual partners as revealed by the Biobank questionnaires. Neither did they find a relationship between any other phrenological characteristics and skull topography. They then put the nail in the coffin by demonstrating that brain gyrification, the folds that appear on the surface of the brain, have no relationship to the shape of the skull. "According to our results," the scientists remarked somewhat tongue in cheek, "a more accurate phrenological bust should be left blank since no regions on the head correlate with any of the faculties that we tested."

While science has now shown that phrenology has no scientific basis, phrenologists do deserve credit of sorts for introducing the idea that different parts of the brain have different functions. Today we know that speech, sight, motor function, and judgment are controlled by different parts of the brain. But all of you who are now probing the back of your neck searching for the Organ of Amativeness, remember that thanks to the Oxford researchers, who admittedly conceived their study over pints at their local pub, phrenology can be confidently labeled a pseudoscience.

OPEN THAT WINDOW!

Molecules are small. Very small. Very, very, very small. So small that with every breath you inhale several sextillion of them. That's a one followed by twenty-one zeros! About 99 percent of these are molecules of oxygen and nitrogen, but that leaves plenty of room for others such as aromas from the kitchen and scents from personal care products, cleaning agents, pine trees, and flowers. Then there are numerous compounds with no odor at all.

Let's consider something as mundane as baking bread. The delightful smell generated is composed of dozens of compounds, 2-acetyl-1-pyrroline, (E)-2-nonenal, methional, and maltol among them. Then there is acrylamide, a carcinogen that has no smell and forms when glucose and the amino acid asparagine, both of which occur naturally in flour, react at a high temperature. Toast that bread, and you get even more acrylamide. If you burn the toast, you will be sniffing furanones and polycyclic hydrocarbons (PAHs). They're recognized carcinogens. And if you try to cover up the smell of the burned toast with an air freshener, you will be inhaling the likes of xylene, dichlorobenzene, and limonene, each of which can be shown to have some sort of toxicity under certain conditions. Of course, just because these chemicals invade our body doesn't mean they cause harm. But neither can we conclude that they don't. The chemical complexity of the air we breathe and its possible consequences on health certainly merit investigation.

Such investigations have been carried out, but mostly on outdoor air. Numerous studies have found that air polluted with significant amounts of carbon monoxide, lead, nitrogen dioxide, ozone, sulphur dioxide, benzene, smoke, or tiny particles from rubber tires raises the risk of heart disease, lung disease, cancer, and even cognitive damage. People living near heavily trafficked areas are affected the most. However, much less is known about indoor air despite concerns about it having a long history.

The effects of inhaled smoke were undoubtedly noted when humans began to heat their caves with fire. Benjamin Franklin was aware of the smoke problem and made sure that the "Franklin stove" he invented produced less smoke than a fireplace. He was also aware of the need for ventilation. John Adams, who once shared a room in a country inn with Franklin, preferred to sleep with windows closed. As Adams recorded in his diary, Franklin objected, and remarked that "The air within this chamber will soon be, and indeed is now worse than that without doors. Open the window and come to bed, and I will convince you." We don't know if Adams was convinced, but today we know that on occasion indoor air can be more polluted than outdoor air.

A series of experiments in 2018 under the framework House Observations of Microbial and Environmental Chemistry (HOMEChem), organized by atmospheric chemist Delphine Farmer and mechanical engineering professor Marina Vance, tackled the indoor air issue. The *house* referred to is at the University of Texas in Austin and was specially built for conducting experiments. It is equipped with all sorts of monitoring equipment to measure chemicals in the air. These include ozone, nitrogen oxides, ammonia, carbon dioxide, particulate matter, and a host of volatile organic compounds (VOCs) that include emissions from cooking, cleaning agents, plants, and human bodies.

Over the period of a month, the researchers and their students cleaned, cooked, sweated, and accumulated data. One experiment involved cooking a Thanksgiving meal with all the trimmings. At one point, the instruments recorded fine particulate matter at a level that if found outdoors would trigger a warning from the Environmental Protection Agency about potentially serious damage to the heart and lungs. Some of these particles came from stir-frying, others from vaporizing deposits that had formed in the oven from previous use. Indeed, self-cleaning ovens that jack up the temperature in an extreme fashion are known to release loads of fine particulate matter. Such particles can be so small, less than a nanometer, that they can even follow the olfactory nerve and pass into the brain. Carbon dioxide levels while the turkey was in the oven reached four thousand parts per million, levels that at least in the short term can impair cognitive function.

There were a number of other interesting findings. Emissions from stir-frying vegetables in teriyaki sauce reacted with vapors from a bleach solution used to mop the floor to produce chloramines, known respiratory tract irritants. When the frying was done on a gas stove, bleach vapors reacted with nitrogen oxides produced by the flames to form nitryl chloride, a substance implicated in smog formation.

Another finding was the presence of hydroxyl radicals that are also involved in smog formation. These are normally produced outdoors when ozone in the air reacts with ultraviolet light in the presence of water vapor. Detecting hydroxyl radicals indoors was a surprise! Apparently, enough ultraviolet light sneaks through windows to produce these radicals that then can react with cooking vapors to produce yet more VOCs.

Human bodily emissions such as rectal gases and squalene, an oily substance produced by sebaceous glands in the skin that prevents the skin from drying out by locking in moisture, were also detected. Numerous chemicals from personal care products such as decamethylcyclopentasiloxane used as a lubricant in creams and shampoos also appeared, as did diethyl phthalate added to perfumes to retard evaporation. Phthalates can also outgas from vinyl flooring where they are used to impart flexibility and are recognized as endocrine disruptors. And we haven't even mentioned the hundreds of compounds found in coffee aroma, the fatty acids in foot odor, the numerous chemicals wafted into the air from the spices used in cooking, or from mold in the bathroom. Furthermore, all these chemicals have the potential of engaging in chemical reactions among themselves, forming yet more substances. Obviously, in indoor air, we are dealing with a chemical stew of immense complexity with essentially unknown health consequences.

Lea Hildebrandt Ruiz, one of the scientists involved in the HOMEChem study also researches outdoor pollution and has monitored air in New Delhi which has perhaps the world's worst air. She noted that fine particulate matter that can reach two hundred and twenty-five micrograms per cubic meter on Delhi's worst days, is still less than the two hundred and eighty micrograms per cubic meter that was detected during the frenzied final hour of cooking the Thanksgiving meal.

What can we take away from all this? Low temperature cooking on an electric rather than a gas stove under a well-functioning ventilation hood is the way to go. And if cooking smells begin to permeate the house, take advice from Benjamin Franklin and open a window. Ben, it seems, was quite an expert on indoor air, even penning an essay on flatulence with a rather provocative title that can be readily Googled.

GIVE A THOUGHT TO URINARY CATHETERS

"It is as flexible as would be expected in a thing of the kind, and I imagine will readily comply with the turns of passage." That was the note sent by Benjamin Franklin to his brother John in 1752 along with his latest invention, a flexible urinary catheter. John suffered from blockage of his urethra by bladder stones, preventing him from emptying his bladder. The solution was to insert a catheter to drain the urine.

The term *catheter* derives from the Greek "to let down," which is exactly what this tube inserted through the penis does. It lets the accumulated urine in the bladder down and out. And with that comes amazing relief, to which anyone who has ever suffered from urinary retention can attest. The pain that comes with this affliction can only be described as indescribable. Take it from one who knows.

But inserting a hollow, straight metal tube into the urethra, as brother John had to do, was not a pleasant experience either. Ben, who had already introduced the lightning rod, bifocal classes, the Franklin stove, and the glass armonica, a musical instrument he said gave him the most pleasure of all his inventions, decided to tackle the problem. He had a silversmith make a springy coil of silver wire that he then coated with tallow to make a flexible catheter. This made insertion much easier, as the flexible tube readily followed the contours of the urethra.

Urinary retention can have several causes, with bladder stones and an enlarged prostate being the most common, explaining why men are more often afflicted. Until roughly 3,500 years ago, if the ailment hit, suffering was inevitable only to be relieved by death. The first treatment, as recorded by the ancient Egyptians, was the insertion of a reed, straw, or bronze tube that, when carefully manipulated, could wend its way into the bladder. A thousand or so years later, Hippocrates wrote of using lead, a soft metal, to make a malleable tube. He also wrote about the possibility of surgical removal of a bladder stone, a procedure first described in the *Sushruta Samhita*, a Sanskrit text written around 600 BC. Hippocrates was wary of such surgery, as can be seen by the famous

Hippocratic Oath including the curious promise, "I will not cut for stone . . . I will leave this operation to be performed by practitioners."

By the first century AD, an S-shaped silver tube was being used for urinary retention as evidenced by such a device being found in the excavation of a doctor's house in Pompei. In the eleventh century, Islamic physician Avicenna used animal skin hardened with white lead to make a somewhat flexible catheter that he lubricated with soft cheese. Then in the sixteenth century, French barber-surgeon Ambroise Pare devised a silver tube with a long, gentle curve for easier insertion. Franklin's contribution came along in the eighteenth century, but how many sufferers other than his brother used it isn't clear.

The introduction of the latex of the rubber tree to Europe from South America resulted in the first attempts to make a truly flexible catheter. The problem was that the rubber would crumble, leaving pieces in the bladder. Not good. Then Charles Goodyear accidentally heated latex with sulphur to produce vulcanized rubber, which had superior properties and allowed French physician Auguste Nélaton to capitalize on the discovery and introduce the Nélaton catheter. By the end of the nineteenth century, such rubber catheters were being produced commercially, but frequent insertion was a problem. Could a catheter be made to remain in place for a longer time? Attempts were made by taping or tying the catheter to the penis, a decidedly uncomfortable method.

French surgeon Jean-Francois Reybard had a better idea. In 1855, he devised a rubber catheter with a tip that had a hole on the side for urine to enter and was also fitted with a small balloon made of sheep intestine that could be inflated with water through a one-way valve. The catheter was thus held in place by the balloon. This was the forerunner of the Foley catheter, introduced in 1935 by Minneapolis urologist Frederick Foley and widely used today when a catheter has to be kept in place for some time, as after prostate surgery or spinal cord injury. It is constructed of a rubber tube with two channels inside, one of which drains urine and the other is tipped with a rubber balloon that is inflated with water after insertion. As with Reybard's device, a valve retains the water in the balloon that then holds the catheter in place.

While the Foley, as all such catheters with this design are now known, solved a problem, it created others. Natural latex is allergenic and can also be toxic to cells. This can result in scarring of the urethra and the formation of strictures that further complicate the release of urine. Catheters coated with silicone rubber or made totally of silicone or polyvinyl chloride (PVC) avoid the issue of allergies, but all catheters are challenged by two other problems. Having a catheter in place often leads to urinary tract infection, as bacteria form biofilms that stick to the material's surface. The catheter can also become encrusted with salts precipitating out from urine, causing it to become blocked. Various approaches attempt to address these problems. Incorporation of silver or other antimicrobials can reduce the chance of infection, and coatings of Teflon, or hydrogels, chemicals that bind water to produce a slippery surface, can prevent encrustation.

The challenge of producing catheters that readily conform to the urethra's anatomy, resist encrustation, and are not plagued by infection issues is very challenging. However, there is no question that science has come a long way from struggling to insert a straw or a straight metal tube into an orifice not designed for such intrusion. Research by modern day Ben Franklins using innovate designs and a range of advanced polymers continues. As for Ben, he, like his brother, came to suffer from bladder stones as the years advanced. He never mentioned whether he made use of the flexible catheter he invented. Maybe he just helped himself by playing his armonica. We now know that sound waves of the right frequency can break up bladder stones. Just a thought.

MY VEGETARIAN GOULASH AND ENDOCRINE DISRUPTORS

My interest in endocrine disruptors was sparked in a most curious fashion. Back in the 1980s, there was a little fruit and vegetable store around the corner from us that also had a deli counter manned by a Czech immigrant who fancied himself as a chef. One day he cooked up a batch of goulash that I thought lacked flavor because he had not used the proper Hungarian paprika. That prompted me to make my own version for him with the right stuff. He liked it! Since I had made a large pot, he started to package the leftovers for sale. As he was doing that, a customer asked if my dish contained meat. Somewhat stunned by the question, I replied, "It's goulash, of course it contains meat." "Too bad," she replied. "I don't eat meat."

"Hmmm," I thought. Concerns about eating meat, along with the benefits of a plant-based diet, were just starting to make headlines at the time, so the prospect of a vegetarian version of my concoction plopped into my mind. But what sort of protein could I use to replace meat? My culinary background certainly did not include tofu, but this soy milk curd seemed like a viable candidate. After some experimentation, I did manage to come up with a reasonable facsimile and I passed the recipe on to the "chef." To my surprise, customers sang its praises! Then one day I happened to be in the store when a customer commented that she would have liked to try "Dr. Joe's Vegetarian Goulash," but as she had been diagnosed with breast cancer, she couldn't. It contained soy, she said, and that had a form of estrogen that she was not allowed since her breast cancer was "estrogen receptor positive." And with that comment my dive into the deep waters of endocrine disruptors began.

Breast cancer has long been recognized as a scourge given that tumors can be visible. As early as 1500 BC, the Egyptian medical text known as the Ebers Papyrus described breast cancer as an incurable disease. Hippocrates, who postulated that health depends on the balance of four humors, namely blood, phlegm, yellow bile, and black bile, speculated that the cause was an excess of black bile since tumors sometimes turn black. Over the years, many other theories were advanced, including Bernardo Ramazzini's thesis in the eighteenth century that nuns had a high incidence of breast cancer due to a lack of sexual activity. German physician Friedrich Hoffman had a different view and suggested that the disease was caused by an excess of such activities. The eighteenth century also introduced the first effective treatment, surgical removal of the tumor. Then with the advent of anesthetics in the nineteenth century, radical mastectomy, along with the removal of lymph nodes, became the standard therapy.

A connection between breast cancer and hormones first surfaced in 1895 when Scottish surgeon George Beatson removed the ovaries from a patient with an ovarian cyst, a common procedure at the time. The patient also happened to have breast cancer, and Beatson noted a shrinkage of her tumor. A clue about this association did not emerge until 1906 when ovarian extracts were shown to stimulate the sexual reproductive cycle, or estrus, in female mammals. The active ingredient in the extract was finally isolated in 1929 and named *estrogen*, deriving from the Greek *oistros* for *mad desire*, and *gennan*, to *produce*. An explanation for why removal of the ovaries resulted in shrinkage of breast tumors now became apparent. Estrogen feeds the tumor!

The mechanism by which this happens was uncovered by organic chemist Elwood Jensen in 1958. Estradiol, the specific estrogen produced by the body, binds to proteins in cells he termed *estrogen receptors*. The receptor-estrogen combo then migrates into the cell's nucleus where it interacts with DNA and causes changes in the expression of specific genes. The result is irregular cell multiplication, the hallmark of cancer.

The discovery of estrogen receptors also suggested a treatment for breast cancer. If these receptors could be occupied by some substance other than estradiol, then binding with estradiol, and hence the irregular cell multiplication could be prevented. That approach worked! Drugs such as tamoxifen bind to estrogen receptors and have become the mainstay of treatment for estrogen receptor positive cancers. Another approach is the use of aromatase inhibitors, drugs that block the activity of aromatase, an enzyme the ovaries use to produce estrogen.

Now we are ready to tackle the soy issue. In the 1930s, Japanese chemists isolated isoflavones, compounds found in soybeans that a decade later were shown by British biochemist Edward Charles Dodds to trigger the same reactions in animals as natural estrogen. This introduced the concept of phytoestrogens, plant-derived compounds with estrogenic activity, and raised the obvious question of whether soy products could "feed" breast tumors. Now, after literally thousands of published studies about soy and isoflavones, we have an answer. Although isoflavones do enhance the proliferation of breast cancer cells in vitro, and can promote estrogen-dependent mammary tumors in rats that have had their ovaries removed, numerous human epidemiological studies have shown that there is no need to be concerned about isoflavones aiding and abetting breast cancer. Indeed, they are likely to do the opposite.

The initial studies to explore the link between soy and breast cancer were prompted by the observation that soy consumption is far greater in Asia than in North America and that the incidence of breast cancer is substantially higher in North America than in Asia. Furthermore, genetics do not seem to be involved since descendants of Asian immigrants to North America take on the local eating pattern and are no longer protected. Add to this case-control studies that compare lifelong diets of breast cancer patients with controls and find that soy consumption, if anything, has a protective effect. The protection seems to be the greatest with significant soy consumption around the age of puberty. There are also a number of prospective studies that have examined soy intake by patients after being diagnosed with breast cancer and found reduced mortality and reduced recurrence with increased consumption.

How then is it that isoflavones, compounds that definitely do bind to estrogen receptors, are not implicated in causing mischief? The answer appears to lie in the relative strengths with which the two substances bind. Isoflavones bind too weakly to activate the receptors, but by occupying the binding sites they prevent them from engaging with estradiol.

Unfortunately, the commercial version of my vegetarian goulash disappeared when the store went out of business. If it were still available, I would have no hesitation in recommending it to anyone. I have no fear of phytoestrogens, chemicals that incidentally are also antioxidants and are found in many fruits and vegetables. To be honest, though, as far as taste goes, my tofu goulash does lack something. Meat.

FROM MICHELANGELO TO RED DYE NO. 3

How do you pivot from Michelangelo painting the ceiling of the Sistine Chapel to looking for cats with goiter in Milan and end up writing about a petition to ban erythrosine, commonly known as Red Dye No. 3? That circuitous route began a few years ago when I was invited to give a talk in Milan, a city I had not visited before. By chance, years earlier, when I was preparing a lecture on hormones I had come across a paper in the *Journal of the Royal Society of Medicine* with the alluring title, "Michelangelo's Divine Goiter." It described an autobiographical sketch of the artist painting the vault of the Sistine Chapel in which he portrayed himself with a swollen neck. The sketch was accompanied by a poem that began with the lines "In this hard toil I've such a goiter grown, Like cats that water drink in Lombardy."

It was in 1508 that Pope Julius II commissioned Michelangelo to paint the ceiling of the Sistine Chapel. This required a scaffold from which the artist could reach up and carry out the difficult task that he claimed was responsible for his goiter, today recognized as a swelling of the thyroid gland. But why the reference to cats in Lombardy? As early as the first century BC, Vitruvius Pollio, a military engineer serving under Julius Caesar, suggested that "swollen throat," which was common in the alpine regions of Italy, was due to drinking river water. He may have been on the right track since water in the area has little iodine and iodine deficiency can lead to goiter. Indeed, after iodine was added to salt in the 1970s, the incidence of goiter in the mountainous areas of Tuscany and Lombardy dropped dramatically.

Milan is the capital of Lombardy, so naturally I looked for cats with goiter. There were plenty of cats, but not one did I see with a swollen neck. It has been proposed that Michelangelo may have been speaking metaphorically, as during his time, peasants in Northern Italy were nicknamed *cats*. The story of Michelangelo's goiter gets even more interesting when the reason for the use of the word *divine* in the title of the paper is revealed. The authors, both pathologists, claim that one of the frescos, *Separation of Light from Darkness* shows the Creator with a goiter. This, they maintain, cannot be an accidental feature since Michelangelo was a perfectionist and had extensive knowledge of anatomy. He even had his own dissecting room at the Church of Santo Spirito in Florence!

But why would Michelangelo have painted the Creator with a goiter? The theory is that he "signed" his Sistine Chapel epic by portraying himself as God in the *Separation of Light from Darkness*, the final panel he painted. An interesting possibility, given that the artist did grow up in Tuscany where goiter was endemic, and the theory would seem to mesh with the sketch he made of himself that clearly shows a swollen neck. A subsequent paper by Italian scientists takes issue with the theory that Michelangelo suffered from goiter since he was only in his thirties when he painted the Sistine Chapel and lived to the ripe old age of eighty-nine without any manifestation of thyroid problems. Their interpretation of the artist's sketch is that the swelling in the neck is actually not a goiter but his larynx that becomes prominent as his neck is stretched while reaching up with the paintbrush.

Now for the Red Dye No. 3 connection. When I was looking into the supposed goiters in the Lombardy cats, I did a search for "thyroid disease in animals." A couple of studies described how an observation of goiter on piglets in Montana and sheep in Michigan prompted experiments that led to the use of iodized salt to prevent goiter in people. But to my surprise, up popped numerous articles that discussed Red Dye No. 3 causing thyroid cancer in rats and possibly in humans. There were also petitions to sign that urged governments to ban the "cancer-causing food dye" and posts that expressed satisfaction with California's proposal to do just that by 2027.

Before going any further, let me say that I have no issue with banning food dyes since these additives serve only a cosmetic purpose and contribute no nutritional value. Indeed, they often serve to attract children to junk foods and have been associated with causing behavioral problems. But how realistic is the cancer connection? I was intrigued enough to look into the original literature that triggered the anti-erythrosine frenzy.

Concerns about erythrosine, a synthetic dye that had been used as a food additive since 1907, were first raised in the 1970s with the realization that its molecule structure resembled that of thyroid hormones including the incorporation of atoms of iodine. Any substance that may have hormone-like activity rings an alarm bell with toxicologists, and as early as 1981 a study in subjects given 25 milligrams of erythrosine a day, ten times the average per capita daily intake, found no effect on thyroid function. Rats first crawled into the picture a year later with a study that showed an increase in the incidence of thyroid tumors in rats fed a diet containing 4 percent erythrosine. That is some ten thousand times what the average North American consumes per day!

A study in 1987 with human subjects showed an increase in thyroid stimulating hormone (TSH) produced by the pituitary gland upon ingestion of erythrosine which could be worrisome since stimulation of excessive activity in the thyroid gland could lead to cancer. But the amount given to subjects that caused an increase was 200 milligrams a day, an unrealistic dose. No increase was seen at a 60-milligram dose, which is still way more than the average exposure. Then in 1988, the trial that spawned the drive to ban erythrosine was published in the *Japanese Journal of Cancer Research.*

Rats were injected with a nitrosamine that is known to cause cancer and some were fed a 4 percent erythrosine diet. After a month, half the thyroid gland was surgically removed in one set of rats. Four months later, all the animals were sacrificed and their thyroid glands examined. The rats that were fed erythrosine, and had half their thyroid removed, presented more tumors in the remaining half-gland than the animals that had not been fed the substance. But, and a big but it is, the rats that were fed erythrosine and had not been subjected to thyroidectomy showed no such effect! Yet, this is the study that is generally quoted when linking Red Dye

No. 3 to cancer. Not exactly what one would call compelling evidence, but then again, when it comes to food dyes, is compelling evidence for a ban really needed? Why take any risk in the absence of benefit?

Incidentally, the first person to produce anatomically accurate drawings of the thyroid gland was Leonardo da Vinci, Michelangelo's contemporary. He too may have a connection to goiter, as it has been suggested that Mona Lisa's yellowing skin, thin hair, and lack of eyebrows is due to hypothyroidism. A hyper stretch, I think.

PRESERVATIVES ARE NOT EVIL

"Let's add some useless synthetic preservatives to our product so that we can increase our expenses, frighten our customers with chemical terms, and possibly even make them sick." I have never heard any food or cosmetic producer utter such a statement. Far from being useless, preservatives help us fight a war we are constantly waging. That war is with molds, fungi, and bacteria, all of which can contaminate our food and cosmetics. These microbes are "natural" and can make us sick. Why then do some consumers seek out products that claim "no preservatives added?" The only possible answer is that they think the risk posed by preservatives is greater than the risk posed by microbial contamination. They are wrong.

Let's take a look at one of the most maligned classes of preservatives, the parabens. These compounds, first introduced in the 1920s, are all derivatives of parahydroxy benzoic acid and are effective against molds, yeasts, and many bacteria even at very low concentrations. They have low allergenicity and find particular use in cosmetics which due to their moisture and fat content, coupled with storage at room temperature, present a particularly hospitable environment for microbial growth. To be sure, not all microbes are dangerous, but some are. Cosmetics can pick up microbes during manufacture or from consumers who repeatedly dip their fingers into a product they apply to their skin. That process can introduce skin microbes that then multiply in the friendly moist environment and present a risk of infection with a subsequent application.

Parabens were cruising along nicely until a 1998 paper published in *Toxicology and Applied Pharmacology* reported that they can bind to estrogen receptors on cells. This raised eyebrows because hormones like estrogen regulate many body functions and can cause problems if levels stray from the normal range. Indeed, irregularities in the development of the uterus were observed in rats exposed to parabens. Another concern was that some breast cancers are estrogen receptor positive, meaning that their growth can be stimulated by exposure to estrogen or possibly by compounds like parabens that bind to estrogen receptors.

Often left out of the discussion is that the potency of parabens to bind to receptors in the rat uterus was some 100,000 times less than that of estradiol produced naturally in the body. Furthermore, we live in a world that is replete with phytoestrogens, compounds found in plants that have estrogenic activity. In this category are isoflavones in soy, lignans in flax, resveratrol in red wine, and coumestrol in spinach and alfalfa sprouts. Phytoestrogens, albeit in small amounts, can be detected in virtually all fruits, vegetables, and legumes; estrogenic activity is not necessarily bad. Phytoestrogens in soy, for example, can bind to estrogen receptors without triggering any activity and at the same time prevent the body's natural estrogen, beta estradiol, from binding. This is desirable in the case of estrogen receptor positive cancers.

In 2002, the pot was further stirred when Japanese researcher Shinshi Oishi published a study that showed decreased sperm production and decreased testosterone levels in male rats fed propyl paraben at doses that approximated the acceptable daily intake for humans. Almost all criticism leveled at parabens since that time, and there has been plenty of it, reference this study as a major reason to avoid parabens. That's despite the publication in 2013 of a study by French researchers using superior methodology, more rats, and higher doses of parabens than Oishi that totally failed to replicate his results. There were no effects on reproductive function!

While parabens' effects on the rat uterus and possible interference with sperm production were of interest to researchers, it was a 2004 study by University of Reading's Dr. Philippa Darbre that ignited the firestorm that swept through the media and brought parabens to the attention of the public. She examined tissue that had been removed from breast tumors in a small group of twenty women and detected trace amounts of parabens. She did not claim that the chemicals caused the disease but pointed out that parabens are used as preservatives in a range of cosmetics applied to the underarm and breast area and that "it had been suggested that regular application of such estrogenic chemicals could influence breast cancer development." Suggestion of course is not the same as demonstration. Much more importantly, Dr. Darbre did not test healthy women to see if they also had these compounds in their breast tissue. Nevertheless, in the eyes of the media, the story became "antiperspirants can cause breast cancer." Never mind that none of the major antiperspirants were preserved with parabens.

Darbre's subsequent experiments revealed that indeed just about everyone has some parabens in their tissues due to the widespread use of these preservatives in various products. While she has not absolved parabens of all blame, Dr. Darbre has gone on record saying that there are hundreds of estrogenic chemicals in the environment, both natural and synthetic, so that singling out of one set of such substances as a guilty party for affecting human health is unrealistic. In spite of its methodological flaws, Dr. Darbre's paper has been frequently referenced with allusions to a cause-and-effect relationship between parabens and breast cancer, one that it certainly did not establish.

In fact, no study has provided any evidence that parabens cause cancer. The International Agency for Research on Cancer (IARC), not known for sweeping any hint of carcinogenicity under the carpet, does not even list parabens as a possible human carcinogen, a group in which it lists some 200 substances. Nevertheless, the "no parabens" claim still attracts consumers. Are they putting themselves at risk of infection? No. Such products contain other preservatives such as DMDM hydantoin, quaternium-15, phenoxyethanol, or imidazolidinyl urea. These have also attracted the ire of some consumer advocacy groups despite numerous studies attesting to their low risk and significant benefit. Then there are preservatives like grapefruit seed extract or tea tree oil that are hyped as being "natural" but have questionable efficacy. Actually, even parabens could be called "natural" since they can be found in tiny amounts in blueberries, carrots, and onions. Of course, the safety or efficacy of a chemical does not depend on whether it was made by a chemist in a lab or Mother Nature in a bush.

When it comes to preservatives, I'll take them over microbes.

THOSE BROWNING APPLES AND ENZYMES

The first chemical reaction to which I remember paying any attention was between polyphenols and the enzyme polyphenol oxidase (PPO). That was long before I knew anything about chemistry. My mother's remedy for any sort of upset stomach was grated apple sprinkled with some sugar. I recall that it didn't take long for the slivers of apple to turn brown.

I didn't give this color change much thought until I started to teach organic chemistry and made a point of emphasizing practical applications. Reactions of a substance with oxygen are a common feature in organic chemistry. The browning of an apple or banana is a typical example and is related to polyphenols, a class of compounds that occur in many plants and are believed to contribute to the "healthy" properties of plant foods. However, if a plant is damaged, polyphenols can react with oxygen and are converted to quinones that then undergo a polymerization reaction to form melanin, a brown pigment. In plants, melanin is believed to protect damaged tissues from further damage by UV light and to deter insects looking for crevices in which to deposit their eggs. Melanin is the same substance that occurs naturally in the human body and is responsible for the color of brown eyes, nipples, dark skin, and hair.

Enzymes are special proteins that act as catalysts, meaning that they speed up chemical reactions without being consumed. PPO is such a catalyst stored in chloroplasts, the organelles in a cell where photosynthesis takes place. Polyphenols on the other hand are stored in cell compartments called vacuoles. When a cell is damaged, the enzymes, polyphenols, and oxygen from the air come together and the browning reaction begins. Grating an apple, or subjecting a banana to cold temperatures, damages cells.

A common experiment suggested to children is to treat a cut-up apple with various liquids to see the effect these have on browning. Inevitably they conclude that lemon juice stops the reaction. Why? Enzymes are sensitive to acidity and tend to function only within a specific pH range. Lemon juice is quite acidic, and, as a bonus, citric acid in the juice can bind metal ions. PPO contains copper ions that are essential for its functioning, and citric acid can pilfer them from the enzyme.

Now we come to an interesting question. Does the browning reaction have an effect on the nutritional quality of plant foods? That is worth exploring since the loss of polyphenols is not desirable. These compounds can neutralize the reactive oxygen species that are byproducts of normal metabolism and have been linked with conditions ranging from cancer to heart disease. For some insight we can look to an interesting study that focused, of all things, on smoothies!

In a paper published in the journal *Food and Function*, researchers reported on the effect of a food with a high content of PPO, such as a banana, on the bioavailability of polyphenols. A specific amount of flavanols, a class of polyphenols, was added to a banana-based smoothie and to a berry-based smoothie. Berries were selected for their low PPO content. Subjects then consumed the smoothies and had their blood and urine tested for flavanol metabolites to determine how much of the flavanol had been absorbed. With the banana-based smoothie, the flavanol metabolites were decreased by about 80 percent relative to the berry-based smoothie. Based on this observation, the researchers suggest that if one wants to have the maximum benefit from polyphenols in a smoothie, it is best to avoid including a banana.

Why did they choose flavanols as a representative of polyphenols in this trial? A clue comes from the acknowledgment that the study was funded by Mars Inc., a chocolate manufacturer. Flavanols are found in cocoa, and Mars has had a long interest in exploring their health benefits based on early studies that demonstrated a reduction in blood pressure. The company has also sponsored the double-blind, placebo-controlled Cocoa Supplement and Multivitamin Outcomes Study (COSMOS), which found a reduction in cardiac deaths in subjects taking a cocoa supplement. It is important to point out that the supplement pills were not made from chocolate, which is an unreliable source of flavanols, but rather from a specially developed cocoa extract. Therefore, this study cannot be used to claim any health benefit for eating chocolate, and indeed the company makes no such claim. But it does have an understandable interest in exploring potential benefits of flavanols and how these may be impacted by other components of a diet such as polyphenol oxidase.

I took particular note of the banana smoothie study because of personal interest. My breakfast every day is avocado toast made with sourdough bread and a bowl of berries, mostly blue, topped with some banana slices and a spoonful of Fiber One cereal. I can make a case for this being healthier than a Danish, or even a bagel with cream cheese. Since blueberries are touted for their polyphenol content, you can appreciate why the study about the PPO activity of bananas caught my eye. But that is all it did. I am not expelling bananas from my berry breakfast. I think in the context of an overall diet, whatever reduction in polyphenol intake occurs by letting my slices of banana mingle with my blueberries is inconsequential.

Let me return to my childhood and those grated apples. Could they really have done something to settle my stomach? Apples contain pectin, a carbohydrate that when mixed with sugar forms a gel. That is the chemistry behind making jams. At one time, it was suggested that pectin via such gel formation can bulk up stools and prevent diarrhea. It was even included along with kaolin, a type of clay, in Kaopectate to treat diarrhea. When it was determined that kaolin alone did the job just as well, pectin was removed from the preparation. There is still a smidgen of evidence that pectin may be of some benefit for diarrhea associated with irritable bowel syndrome. In this case, it may be acting as a prebiotic, serving as food for bifidobacteria, microbes that then multiply and squeeze out bacteria that can cause diarrhea.

I think I can conclude that the grated apples were unlikely to have had any sort of therapeutic effect other than providing small bits of apple that are easy to digest. But nevertheless, likely out of nostalgia, when I have some sort of digestive issue, I'll go for the grated apple. But I do squeeze lemon juice on it. And thanks to chemistry, I even know why. And now, so do you.

WHEN IT COMES TO HORMONES, SEX TOYS ARE NOT SO SEXY

Be careful, sex toys may be leaching microplastics and phthalates! Back in the 1990s, when I started writing these little columns, I could never have imagined starting with such a statement. While I would have been familiar with phthalates as plasticizers, chemicals added to plastics to impart flexibility, sex toys would not have been on my radar, and the term *microplastic* was not yet in anyone's vocabulary.

How things have changed! Today I'm willing to tackle a paper published about sex toys and toxins. After all, sex toys are widely advertised online, numerous articles discuss the risks posed by discarded plastics producing microplastics, and phthalates have become villains as endocrine disruptors. That term was coined in 1991 by a diverse group of twenty-one experts who had assembled at the Wingspread Conference Center in Racine, Wisconsin, to assess the possible interference of industrial and agricultural chemicals with hormonal activity in wildlife and humans. *Endocrine* means *secreting internally* and refers to the discharge of hormones by various organs and glands into the bloodstream. Endocrine disruptors interfere with this process.

In 1962, Rachel Carson's classic book, *Silent Spring*, had called attention to the harmful effects of pesticides such as DDT's ability to interfere with the reproductive capabilities of birds. Carson's epic stimulated extensive research into chemicals released into the environment and inspired biologist Theodora Colborn to organize the Wingspread Conference. She had herself observed abnormalities in wildlife in the Great Lakes region with male birds growing ovarian tissue and fish having both male and female reproductive organs.

The scientists at the conference were aware of, and discussed extensively, the consequences of the use of diethylstilbestrol (DES), a synthetic estrogen mimic that had been prescribed between 1940 and 1971 to prevent miscarriage. Daughters born to these women showed an increased risk of a rare form of vaginal cancer, exhibited altered immune response, and had reduced fertility later in life. These effects were similar to those found in wildlife and in laboratory animals exposed to industrial chemicals with estrogen-like effects, suggesting that such chemicals can also pose a risk to humans.

Tufts University scientist Dr. Ana Soto was invited to the Wingspread Conference because of a remarkable observation she made that had steered her research into the field of endocrine disruptors. Dr. Soto's group had been studying the effects of estrogen on the multiplication of breast cells and discovered that a solution that had no estrogen in it still caused the cells to react as if it did contain some of the hormone. After months of experiments Soto discovered that the solution was leaching nonylphenol from the plastic test tube in which it was stored. Nonylphenol is an antioxidant added to some plastics to prevent breakdown, and, as Dr. Soto discovered, it has estrogen-like properties. Her concern was further elevated on learning that the compound is also used in the formulation of detergents and spermicidal creams. That was enough motivation to start a search for other hormone-like chemicals to which people may be exposed. The result was a landmark joint publication with Theo Colborn and University of Missouri professor Frederick vom Saal in 1993 that listed a number of endocrine disruptors including various herbicides, insecticides, and industrial chemicals such as PCBs and phthalates!

The three decades since that paper was published have seen an explosion of research into endocrine disruptors. Additions to the list include bisphenol A, used to produce polycarbonate plastics and epoxy linings in canned foods; perfluoroalkyl substances (PFAS) that appear in a myriad products thanks to their oil- and water-resistant properties; various flame retardants; the preservative parabens; the sunscreen oxybenzone; and a host of naturally occurring phytoestrogens found in tea, berries, fruits, vegetables, and soybeans. These substances can either mimic or block the activity of the body's hormones.

The phytoestrogens in general have positive effects, with some such as isoflavones in soy showing breast cancer preventative activity. However, the industrially produced endocrine disruptors have been linked with obesity, thyroid disorders, type 2 diabetes, and cancer. Most of the studies have used either animals or cell cultures, although some have tested human blood for the presence of endocrine disruptors and found a correlation with various ailments. Early onset puberty and falling sperm counts have also been associated with an increase in endocrine disruptors in the environment. Associations, of course, cannot prove a cause-and-effect relationship, but scrutinizing the literally thousands of studies on endocrine disruptors leads to the conclusion that efforts to reduce exposure to these chemicals are well founded.

This now brings us back to the sex toys. Researchers were interested in exploring any chemical risks they may present given that about half of heterosexual adults and over seventy percent of the LGBTQ population report using or having used such toys. Since these devices make close contact with intimate body parts, transfer of chemicals they may release warrants investigation. The products were subjected to a machine that abrades their surface to release microplastics which were then collected and extracted with a solvent. The solution was then analyzed by gas chromatography/mass spectrometry. All samples contained phthalates in concentrations exceeding European and U.S. regulatory standards for similar products such as the PVC duckies that children may put into their mouths.

While this study is intriguing, there are a number of caveats. The use of an abrasion machine is not exactly analogous to the friction that is generated by conventional use. And without measuring blood levels, it is impossible to know if any phthalates end up in the bloodstream. For what it is worth, and I can't believe I'm actually imparting this information, the release of microplastics was, from most to least, stubby nubby anal toy, anal beads, rabbit vibrator, and an external vibrator. The authors do not claim to evaluate risk in any fashion but aim to increase awareness of the presence of potential endocrine disruptors in such products and the need for regulatory agencies to examine the issue.

As I was exploring this stimulating topic, I happened to come across an ad for "Sex Oil" from Gwyneth Paltrow's Goop Company. It was described as a "plush-oil jelly that sinks into the skin and adds just the right amount of slip and is great for all body parts" (which are actually named). The ingredients list a dozen seed oils and various "time-tested aphrodisiac botanicals." I wonder what sort of safety studies Gwyneth has done.

WAR AND CHEMOTHERAPY

The air above the harbor of Bari, Italy, was thick with the fumes from explosions and flames were leaping from ship to ship. Sailors had no choice but to plunge into the waters that were already being covered with a thick layer of burning oil. It was December 2, 1943, and a hundred and five German bombers had just targeted the thirty-one Allied ships that were unloading their cargo of tanks, aviation fuel, ammunition, and other supplies needed to drive the Nazis out of Italy. One of those ships, the American vessel *John Harvey*, was hit directly and discharged its secret cargo of mustard gas into the water, exposing the seamen to the toxic substance as they struggled towards the shore. Discovering the damaging effect of the mustard gas on the white blood cells of the survivors is said to have launched the era of cancer chemotherapy. As is almost always the case, the real story is more complicated than the simplified version commonly presented.

Mustard gas is not actually a gas but a volatile liquid that was first synthesized in 1860 by British chemist Frederick Guthrie who had studied under famed professor Robert Bunsen at the University of Heidelberg. Guthrie combined sulphur dichloride with ethylene and noted that it caused blisters on exposed skin. When the First World War erupted, mustard gas became a candidate for battlefield use and was introduced in 1917 by the Germans under the name *yperite* at the Third Battle of Ypres. Yperite reeked of the smell of garlic or horseradish and had a yellow tinge, accounting for it being termed *mustard gas*. The Allies retaliated in kind, with gas warfare proving to be so terrifying that in 1925 the Geneva Convention banned the use of chemical weapons.

As the Second World War loomed, suspicious that the Germans may not play by the rules, the U.S. initiated a secret research program on the effects of mustard gas at Yale University. One effect of the chemical was already known from autopsies carried out in 1919 on the victims of mustard gas by Dr. Edward Bell Krumbhaar who served as a medical officer with the American forces in France during the Great War. He had documented an unusually low white blood cell count, suggesting that mustard gas affected cell production in the bone marrow.

Drs. Alfred Gilman and Louis Goodman were the Yale researchers charged with investigating mustard gas and focused on a version in which the sulphur atom was replaced by nitrogen since it was believed the Germans were working on such nitrogen mustards that were more toxic than the original mustard gas. Gilman and Goodman began by injecting small amounts into rabbits, triggering jibes by colleagues who opined that the enemy was not likely to attack with hypodermic needles.

The rabbits' bone marrow, spleen, and lymph nodes, the so-called lymphoid tissues, were the most affected, suggesting that tumors in these tissues may be especially susceptible to the cell-damaging effects of the nitrogen mustard. Next step was to implant a lymphoid tumor cell into a mouse and investigate whether nitrogen mustard had an effect on the tumor that grew. It did! The tumor regressed remarkably!

These results were encouraging enough to consider a therapeutic trial in a human. In December of 1942, surgeon Gustav Lindskog at Yale supervised the treatment of a patient who had multiple lymphoid tumors. The results were as dramatic as those seen in the mouse! After ten days of treatment, the tumors had receded to the extent that they were no longer palpable. Unfortunately, the effect was not long-lasting. A tumor regenerated in the bone marrow, a second treatment had minimal effects, and a third had none. Treatment was tried on five additional patients with the same final result. In June of 1943, the Yale research group was disbanded, but there had been proof of concept; nitrogen mustard had the ability to stop cell division in tumors.

The attack on Bari came in December of that year. More than a thousand American and British servicemen were killed immediately and hundreds ended up in hospital. Their blisters, burning sensations, and respiratory problems suggested exposure to some sort of toxin, and doctors wondered if the Germans had used a poison gas. The Americans dispatched Lt. Col. Stewart Francis Alexander, a physician who had special training in chemical warfare and had actually experimented with nitrogen mustard on rabbits. When he saw the low white blood cell counts and blisters in the hospitalized victims at Bari, he immediately suspected mustard gas but was puzzled by the fact that the men who had jumped into the water were the most severely affected. This was not consistent with an aerial attack, which would have affected more people including the civilian population of the town.

Dr. Alexander knew that the U.S. had supplies of mustard gas and now suspected that one of the ships may have carried such a cargo. He was repeatedly informed by both British and American authorities that none of the ships had transported chemical weapons, with Churchill declaring that if the doctor thinks his patients were exposed to mustard gas, he should reexamine them. The Allies were reluctant to admit stocking mustard gas for fear that this would give the Germans justification to use chemical weapons. Alexander's suspicions would eventually prove to be correct: the *John Harvey* had indeed been carrying bombs loaded with mustard gas. The men who had sought refuge in water had the most serious injuries with the chemical leaching through their skin from the wet clothes that had been worn for hours. Dr. Alexander's report of mustard gas injuries was immediately classified, and the official secrecy surrounding the Bari disaster continued for decades.

But armed with the report of the low white blood cell count in the victims, and knowledge about the Yale experiments, Dr. Cornelius Rhoads, chief of the Medical Division of the Chemical Warfare Service, went in search of funding for chemotherapy. He managed to persuade General Motors executives Alfred Sloan and Charles Kettering to create the Sloan Kettering Research Institute for Cancer. The mechanism by which mustard gases cause their effects, namely by attacking DNA, was identified, opening the way to more effective and safer alkylating agents to treat cancer. The Bari disaster did take a step in the march towards effective cancer treatments, but it is incorrect to claim that it resulted in the serendipitous discovery of chemotherapy.

THE LONG HISTORY OF PERFORMANCE ENHANCING DRUGS

The first time I ever heard of performance enhancing substances in sports was in 1960. I was glued to the television set watching the Winter Olympics from Squaw Valley and was stunned when the U.S. defeated Canada, then the Soviets, to advance to the hockey final with Czechoslovakia. The Americans were behind 4–3 after two periods and incredibly scored six goals in the third to win 9–4 to capture the gold.

After the game, a captivating story emerged about one of the Soviet players going into the U.S. dressing room after the second period and advising the players to breathe extra oxygen, apparently a common Soviet practice at the time. The result was the first U.S. "Miracle on Ice," twenty years before the more celebrated one at the Lake Placid Olympics.

Inhaling extra oxygen makes scientific sense since oxygen is needed by muscle cells to produce energy. Breathing pure oxygen can indeed elevate the amount of oxygen delivered to muscles by hemoglobin in red blood cells, but the effect is temporary. Training at higher altitude, where each breath contains less oxygen, is more effective. The kidneys react by churning out more erythropoietin (EPO), a hormone that prompts the bone marrow to increase its production of red blood cells. The more red blood cells present, the more oxygen is available to muscles. The benefits of high-altitude training were brought home at the 1968 Olympics in Mexico City which is located some 7,000 feet above sea level. In preparation, many athletes trained at high altitudes and the effects were reflected in their performance. The runners ran faster, the jumpers jumped farther and higher.

After the Mexico games, some athletes began to wonder if there was an even better way to increase the number of red blood cells that carry oxygen. Why not just withdraw some of your blood, store it in a bottle, and infuse it before an event to increase your red blood cell count? That worked! The concept of blood doping was born and embraced, particularly by cyclists. Then in the 1990s, scientists isolated the gene that codes for the production of EPO and inserted it into bacteria that then multiplied to produce copious amounts of EPO. This was great for patients who suffered from anemia or who were on dialysis because of improperly functioning kidneys. But it didn't take long for athletes to figure that they had been handed an easy method to improve performance. Cheaters got away with injecting EPO until a method of detection was developed for the 2000 Sydney Olympics.

Although the use of EPO was well known within sporting circles, it was Oprah Winfrey's interview with Lance Armstrong in 2013 that brought the matter to the public's attention. Armstrong had won cycling's Tour de France, probably the world's most demanding athletic competition, an incredible seven times! After years of repeated denials, he finally admitted that he had blood doped, injected EPO, used human growth hormone, testosterone, and diuretics. Furthermore, not only had cycling's icon cheated, he ran a veritable empire that provided drugs to others. Armstrong's fall from grace was quick and brutal. He was stripped of all his titles and went from being one of the world's most admired athletes to the most reviled.

Although Lance Armstrong may be the poster boy for cheating in sports, he is far from unique. Athletes have a long history of using performance enhancing substances. Competitors in the ancient Greek Olympics swigged poppy juice, French cyclists in the late nineteenth century drank a concoction of wine and cocaine known as Vin Mariani, and at the 1904 Olympics in St. Louis, Thomas Hicks, the winner of the marathon, got a boost from strychnine. By the 1950s, amphetamines had made it into locker rooms. In 1970, New York Yankee pitcher Jim Bouton in his classic book *Ball Four* described the widespread use of amphetamines, known as "greenies" in baseball. Players who didn't use these stimulants were said to be "playing naked."

Baseball players were not the only ones abusing amphetamines. In 1967, British cyclist Tommy Simpson died during a stage of the Tour de France and was found to have two tubes of amphetamines in his jersey. Simpson's death precipitated testing for drugs at the 1968 Olympics, although tests for steroids were not yet available. Testosterone, the steroidal male hormone that can help build muscle was rumored to have been used as early as 1954 by Soviet weightlifters, but it was only at the 1976 Montreal Olympics that reliable tests for the detection of steroids were introduced. Then in 1988, Canadians felt a collective shame when Ben Johnson was stripped of his 100-meter sprint gold medal for having been propelled by stanozolol, an anabolic steroid.

After the Ben Johnson fiasco, baseball commissioner Fay Vincent sent a memo to all teams informing them that the use of controlled drugs, including steroids, is prohibited, but it had little impact given that no testing was implemented. Then a 2002 article in *Sports Illustrated* that featured an interview with former Most Valuable Player Ken Caminiti blew the lid off steroid abuse in baseball. Games, Caminiti explained, were won and lost not so much by skill, but as to who had the better chemist. With no testing, it was a drug-laced free-for-all. He himself had used steroids that he claimed made him feel like Superman. Players were also injecting so much human growth hormone that, along with their muscles, head sizes also increased to the extent that their helmets would not fit properly.

Why was the use of steroids so attractive? Because they really do increase strength! A sixty home run season is a landmark achievement in baseball and in the 147-year history of the game has only happened nine times, six of which were in the years between 1998 and 2001, at the height of what was known as the "Steroid Era." In those four years, Sammy Sosa hit over sixty home runs three times, Mark McGwire twice, and Barry Bonds set the all-time record in 2001 with seventy-three homers. All three had used steroids! The only other players to hit sixty or more home runs were Babe Ruth, Roger Maris, and Aaron Judge, all of whom were "clean." Prompted by the *Sports Illustrated* revelation, baseball finally instituted testing of players in 2003. Just a year later, the danger of taking steroids was underlined with the death of Caminiti at age forty-one from a heart attack with drugs being determined to have been a factor.

Testing has reduced the use of detectable steroids. There is, however, a constant race between doping detector organizations such as the World Anti Doping Agency founded in 1999 with Canadian swimming champion and former McGill University chancellor Dick Pound as its first president, and clandestine chemists who strive to come up with designer steroids that can escape detection and help athletes in their quest to be faster and stronger at whatever cost. It is hard to know who is ahead in that race.

Now, back to the miracle of Squaw Valley. The six goals scored by the Americans in the third period were all scored by players who had not taken oxygen! Sport does not need performance enhancing substances to be exciting!

THE SEARCH FOR THE MAGIC BULLET

Kill the invader; don't kill the patient. That's what Paul Ehrlich, often called the father of immunology, had in mind when he coined the term *magic bullet* back in 1907. He was talking about a hypothetical drug that could target a microbe or a toxin and destroy it without any collateral damage to the surrounding environment. Today, there are bullets that may not exactly be magical, but they are not hypothetical. We call them *monoclonal antibodies.*

Lady Mary Wortley Montagu arrived in Istanbul in 1716 as the wife of the British Ambassador to the Ottoman Empire. She was not a doctor nor a scientist but nevertheless played a pivotal role in the history of immunology. That's the study of the protective mechanisms the body uses to fend off attack by foreign substances such as bacteria, fungi, parasites, viruses, cancer cells, and toxins, collectively called *antigens.* Lady Montagu had lost a brother to smallpox and had survived an infection herself, of which she was constantly reminded by the scars on her face. No wonder then that she took an interest in a practice she encountered in Istanbul that was supposed to offer protection against this scourge. Pus taken from a blister of a person suffering a mild case of smallpox was scratched into the arm or leg of an uninfected person. Should this person be subsequently infected, the disease would be mild.

Lady Montagu was so impressed she had her son inoculated and on her return to England enthusiastically promoted variolation, as the procedure was called. It was not perfect since the inoculated individual could still infect others and sometimes would contract a serious case of the disease. Edward Jenner sought to improve the procedure and discovered that inoculation with pus taken from the blister of a person infected with cowpox, a much less severe disease, would also protect against smallpox. This was the first *vaccine*, the term deriving from the Latin for *cow*. At the time there was no knowledge of microbes, such as the virus that causes smallpox, and certainly nothing was known about the mechanism that resulted in immunization.

The situation began to be clarified when German physician Robert Koch isolated the bacteria that cause anthrax and Louis Pasteur created a vaccine with an attenuated form of the bacterium. Although the work of Pasteur and Koch established the germ theory of disease, there was little understanding of how immunity was conferred in terms of chemistry. Then, working under Koch, Emil von Behring confronted diphtheria, a bacterial disease that at the time was known as the "Strangling Angel of Children." It caused swelling of the throat to the extent that breathing was impaired. Von Behring made the remarkable discovery that the disease could be cured by injection with serum from a horse that had been inoculated with the diphtheria-causing bacteria. Paul Ehrlich then worked out a method to standardize extracts of the serum, and deaths from diphtheria plummeted.

It was at this point that Ehrlich ventured a theory about what was going on. The horse serum contained a chemical, an antitoxin, that neutralized the toxin produced by bacteria. He theorized that white blood cells feature a large number of potential antitoxins on their surface, and when one of them engages the toxin, it latches on to it and prevents it from doing any damage. Ehrlich called these antitoxins *receptors for toxins* and visualized the toxins fitting into them much like a key fits a lock. When such a fit occurred, the cell produced more receptors, some of which would be released into the bloodstream and hunt down more of the toxin molecules. He called these liberated receptors *antibodies* and described them as *magic bullets*, since they would interact only with the specific substance that caused them to be produced in the first place. Amazingly, Ehrlich was essentially bang on with his description of antibodies.

Of course, much has been learned about antibodies since Ehrlich first floated the magic bullet theory. It is now clear that a natural infection, or being vaccinated, activates many B cells, each of which secretes antibodies that bind to the antigen to different extents. These antibodies come from many different B cells and are said to be polyclonal. A major thrust in immunological research these days is to find a laboratory method to select a specific B cell that makes only a desired antibody. Since they come from only one type of cell, they are monoclonal. These antibodies can then be injected into the body and will recognize and bind to the specific antigen that was used to produce them. That may be a protein on the surface of a cancer cell or one of the body's many naturally occurring chemicals that can cause inflammation.

In 1975, immunologists Georges Köhler and César Milstein made a breakthrough. They injected an antigen into a mouse and then extracted the B cells from the animal's spleen that were specific for the production of the antibody for that antigen. These cells were then fused with a type of cancer cell to form hybridoma cells. They inherited the B cell's ability to produce the antibody, and as is the characteristic for cancer cells, kept multiplying and generating monoclonal antibodies in sufficient amounts to be isolated.

One of the most successful applications of this technology was by immunologist Jan Vilcek of New York University who developed infliximab (Remicade), a monoclonal antibody that recognizes and neutralizes tumor necrosis factor alpha. This molecule regulates inflammation and can cause various autoimmune diseases such as Crohn's or rheumatoid arthritis when it goes out of kilter. Infliximab has proven to be extremely effective at inactivating TNF-alpha, earning billions of dollars for Janssen Biotech and a fortune for Dr. Vilcek, who has used most of the money for philanthropic purposes.

Monoclonal antibodies such as bebtelovimab that target the spike protein of the Sars-CoV-2 virus have been developed, as well as a number of monoclonal antibodies for the treatment of cancer. These "biologicals" have completely changed the approach towards drug development. Historically, most drugs, morphine, digitalis, quinine, aspirin, and penicillin being typical examples, were discovered through trial and error or lucky accidents. Monoclonals on the other hand are based on an understanding of the molecular basis of disease and are designed to selectively target causative factors. They are Paul Ehrlich's magic bullets realized. More or less.

KIMCHI IN SPACE

When the Soyuz rocket blasted off from the Baikonur Cosmodrome in Kazakhstan in 2008 to rendezvous with the International Space Station, it carried two Russian cosmonauts and the first ever South Korean astronaut, engineer Soyeon Yi. Along for the ride was a special batch of kimchi, the traditional fermented Korean national dish that would remind her of home. "If a Korean goes into space, kimchi must go there too," said Kim Sung Soo, a Korea Food Research Institute scientist.

The basic ingredient in kimchi is napa cabbage, which is combined with various vegetables and spices. A typical recipe includes red pepper powder, garlic, ginger, radishes, anchovy juice, sugar, and green onion in addition to the brined cabbage. When allowed to stand in a covered container, the naturally occurring bacteria on the vegetables, mostly of the *Lactobacillus* variety, convert carbohydrates into lactic acid that serves as a preservative by preventing the growth of other bacteria. While there are many varieties of kimchi depending on the specific vegetables and spices used, they all tend to have a potent fragrance, mostly due to sulphur compounds released from the garlic and ginger due to bacterial action.

Kimchi is regarded as both a probiotic and a prebiotic food. Probiotics are bacteria that are believed to be beneficial in terms of health, and prebiotics are food components such as fiber that serve as nutrients for the probiotic bacteria. There was concern that in the Space Station the kimchi would be exposed to cosmic radiation and its bacteria would mutate into a dangerous form. Continued fermentation also produces carbon dioxide, potentially bursting the container and spewing contents all over. Bits of fermented smelly vegetables floating around the interior of the Station were not a desirable prospect.

The Korean Atomic Energy Research Institute rose to the challenge of developing safe "space kimchi" by resorting to irradiation, a method of preservation that exposes food to gamma rays or electron beams to kill bacteria. The kimchi that traveled into space contained no live bacteria, so there was no concern about continued fermentation. As a bonus, the pungent smells were also reduced. While Dr. Yi admitted that the texture and taste of the space kimchi were not quite up to its Earthly counterparts, nevertheless, thanks to its spiciness, it was enjoyed by her international colleagues. Taste buds do not function the same way in a low gravity situation, and astronauts have often complained about food tasting bland.

It turns out that the taste of kimchi depends not only on its components, but also on the container in which it is made. Fermented vegetables became a staple in Korea over a thousand years ago with the observation that they would last longer than fresh vegetables. Of course, it wasn't known at the time that this was due to the lactic acid produced by bacteria creating an acidic environment uninhabitable for most other microorganisms. Traditionally, kimchi was fermented in clay pots called onggis that were buried in the ground for several weeks to maintain a consistent temperature and provide optimal conditions for fermentation. Today, kimchi is mass produced with the fermentation carried out in glass, steel, or plastic vessels. Kimchi aficionados claim that fermentation in onggis produces a superior product, and modern science seems to back up that view.

Researchers in Korea fermented kimchi for three to four weeks at 4 degrees Celsius in glass, steel, polyethylene, and polypropylene containers and compared the results with kimchi fermented in traditional onggis. A panel of tasters judged both the taste and texture of the traditional kimchi to be superior, the latter being evaluated in terms of the "springiness" of the cabbage. Lactic acid bacteria proliferated more readily in the onggis and crowded out putrefactive bacteria, resulting in a longer shelf life for the kimchi.

A test for kimchi's ability to neutralize free radicals, a measure of antioxidant potential, was also performed and revealed that onggi-fermented kimchi had higher radical-scavenging activity. Numerous studies have associated various phytochemicals (compounds found in plants), such as polyphenols that are plentiful in fruits and vegetables, with a reduced incidence of cancer. Therefore, there was interest in exploring whether kimchi extract has an effect on the proliferation of cancer cells. Once more, kimchi fermented in onggi proved to be superior, having a higher antiproliferative effect on human colon cancer cells.

The Korean scientists attributed the noted benefits to the porosity of the clay container. A rapid buildup of carbon dioxide suffocates lactic acid bacteria but the porosity of onggis allows the carbon dioxide to pass through as it is formed. By contrast, the other containers are far less permeable, and unless they are periodically "burped" by lifting the cover, the gas buildup impairs the multiplication of the lactic acid bacteria. The fewer such bacteria, the less antioxidant and anticancer compounds are produced, and the smaller the chance that once the bacteria are ingested they will crowd out disease-causing bacteria in the gut. Indeed, one of the reasons claimed for the health benefits of fermented foods is that they will favorably alter the balance of bacteria in our microbiome.

Does all of this mean that space kimchi is less healthy? For astronauts that question is irrelevant because in any case only small amounts would be consumed for a limited time. But what about the rest of us? Should we be increasing our intake of fermented foods? Here things become complicated. While the microbiome, the diverse families of microbes that inhabit our gut, is an area of intense study, we don't really know what an ideal bacterial population is and how it can be affected by diet. Koreans, who consume kimchi at almost every meal, do have one of the longest life expectancies in the world, but they also have a high rate of stomach cancer, possibly due to the high salt content of kimchi.

Kimchi is not a superfood, no food is. But including fermented foods such as kefir, yogurt, kombucha, miso, tempeh, and, yes, kimchi in the diet is a good idea. Unfortunately, having a fish allergy, I shy away from tasting kimchi because of the anchovies, although I have had some in a Korean restaurant before I knew about the anchovy content and nothing happened. While I think that there is great adventure to be had by trying the dozens of varieties of commercial kimchi available in our grocery stores, I think I will stick to sauerkraut which, like kimchi, is fermented cabbage and has loads of lactic acid producing bacteria. Somewhat surprisingly, there is no record of German or Polish astronauts taking this delicacy into space.

ASPARTAME'S COMPLICATED REPUTATION

Aspartame gives me a headache. Not literally. Figuratively. Ever since the introduction of this artificial sweetener back in 1981 I have followed the skirmishes in the scientific literature about its safety and have probably been asked more questions about this chemical than any other. Those skirmishes are now set to erupt into a large-scale battle as the International Agency for Research on Cancer (IARC), an arm of the World Health Organization, gets ready to declare that aspartame is a possible human carcinogen. Yikes!

This is not the first time that aspartame has been skewered with accusations of carcinogenicity, but a formal listing by IARC strikes a different chord. Consumer fears will surely hit a high note and panic will flare in the boardrooms of the numerous food and beverage companies that aim to sweeten people's lives without plying them with sugar.

Back in 1965, G.D. Searle Company chemist Jim Schlatter was carrying out research on gastric ulcers and certainly did not have artificial sweeteners in mind. He knew that entry of food into the stomach stimulated the secretion of gastrin, a hormone that triggers the production of gastric acid. At the time, the common belief was that ulcers were caused by excess stomach acid, and Schlatter was interested in finding a drug that could inactivate gastrin.

In the course of this research, Schlatter synthesized some model compounds that incorporated certain features of the hormone. One day, after licking his finger to pick up a sheet of paper, he noticed a sweet taste, which he eventually traced to the aspartylphenylalanine methyl ester he had just synthesized. Little did Schlatter dream that within twenty years his discovery would be netting the company a billion dollars of profit a year! And he most certainly never imagined that an IARC classification would embroil his sweet crystals in a bitter controversy.

IARC catalogues chemicals about which questions of carcinogenicity have been raised as "carcinogenic to humans," "probably carcinogenic to humans," or "possibly carcinogenic to humans." Tobacco, asbestos, alcohol, and sunlight are known carcinogens, while red meat, extremely hot beverages, lead, and hairdressing as a profession are ranked as probable carcinogens. Pickled vegetables, gasoline vapor, extracts of ginkgo biloba, and aloe vera are listed as possibly carcinogenic. It is into that latter category that aspartame is to be stuffed.

Now for the crux of the matter. IARC determinations are based on an analysis of hazard, not risk! Hazard is the innate property of a substance or process to cause cancer and cannot be changed. Risk, on the other hand, is an evaluation of whether a hazardous substance causes harm under real-life conditions. It can be altered! Ultraviolet light from the sun is definitely a hazard, but the risk from its effects can be reduced by wearing sunscreen or just staying out of the sun.

Listing aspartame as possibly carcinogenic means that there is some evidence that under some conditions, be it in cells in a petri dish or in some animal feeding study, the chemical was shown to cause some form of cancer. A couple of studies in mice have linked aspartame to cancer, but their methodologies have been widely criticized. Even if feeding large doses to mice does indeed cause cancer, one cannot assume that the smaller doses to which humans are exposed will do the same. But still that is enough for IARC to hang a *possibly carcinogenic* sign on the neck of aspartame.

The most often cited human study linking aspartame to cancer is the NutriNet-Santé study in France that followed some 100,000 people who periodically filled out dietary questionnaires over twelve years. The researchers reported an increased risk of cancer associated with the consumption of artificial sweeteners, especially aspartame, acesulfame potassium, and sucralose. How large an increase? For every thousand subjects who had a high intake of aspartame, there were thirty-three cases of cancer detected. For those who consumed no aspartame, there were thirty-one cases. That is a tiny difference of 0.2 percent! And of course, such observational studies cannot prove a cause-and-effect relationship. Maybe consumers of artificial sweeteners eat more processed foods or fewer vegetables. Furthermore, dietary questionnaires are notoriously unreliable. It is also difficult to see how different artificial sweeteners with totally different molecular structures can have the same effect.

In the tit for tat world of aspartame studies, we can refer to the Nurses' Health Study and Health Professionals Follow-Up Study in the U.S. that also followed more than 100,000 subjects and found no association with cancer. In yet another study with a similar number of participants, American Cancer Society researchers were unable to detect any link between artificial sweeteners and cancer. But a study of some 4,000 cancer patients matched against controls in Spain did reveal a small increased risk of cancer with high consumption of aspartame, but only in diabetics. And so it goes. Overall, though, the number of studies that find no link to cancer dramatically outnumber those that do.

Leaving the cancer issue aside, there are other concerns about aspartame and its brethren sweeteners. They can affect our microbiome, the composition of bacteria in our gut! Such changes are now suspected of playing a crucial role in numerous conditions including digestive problems, obesity, heart disease, type 2 diabetes, and even depression.

There are other concerns as well. A sweet taste in the mouth prompts the body to expect sugar and release insulin. But if the sugar does not arrive, blood glucose drops and that creates hunger, especially for carbohydrates. Maybe that is why studies indicate that replacing sugar with non-caloric sweeteners fails to control weight in the long term. Indeed, long-term use is associated with an increased risk of obesity, type 2 diabetes, and cardiovascular disease. Of course, there is always the possibility of reverse causation, meaning that people suffering from such conditions choose to use sweeteners hoping to remedy the problem.

We will have to wait and see on what basis IARC proposes to list aspartame as a possible carcinogen. In the meantime, there is enough evidence, aside from carcinogenicity, to suggest that the benefits of non-caloric sweeteners have been overhyped and that efforts should be made to cut down both on sugar and sweeteners. Nothing wrong with drinking your coffee black or having water with your meal. And berries do a great job in sweetening yogurt.

SKELETONS IN THE ALCOHOL CABINET

"Moderation in everything" is a popular slogan. It seems to go hand in hand with the cornerstone of toxicology, namely Paracelsus's famous dictum that "only the dose makes the poison." We use it justify eating just a few potato chips instead of the whole bag, a spoonful of ice cream instead of the whole container, and a glass of wine instead of half a bottle. Ah, the wine. There may be a snag here. It's the alcohol. A known carcinogen. The International Agency for Research on Cancer (IARC) places ethanol in "Group 1," reserved for substances such as tobacco, asbestos, and radioactive materials that are "known to cause cancer in humans," and the World Health Organization has concluded that the only safe amount of alcohol to consume is zero. No "moderation" here. Not a comforting thought.

But what about all those studies, and there have been many, that have documented a *J* shaped curve when alcohol consumption is plotted against cardiovascular disease? Compared with zero alcohol, risk decreases until the consumption of 10 grams of alcohol a day is reached, roughly that found in a glass of wine, and then increases sharply.

These studies have comforted many who enjoy having that glass of wine with dinner and even prompted some to take up the habit. Various studies have buttressed the argument for red wine by pointing out that it contains numerous polyphenols, resveratrol in particular, that act as antioxidants and prevent the formation of oxidized cholesterol, the villain implicated in the formation of artery blocking plaque. Others point out that alcohol can raise levels of HDL, the so-called "good cholesterol," reduce levels of inflammatory proteins, release vasodilating nitric oxide, and impair the ability of platelets to aggregate and form blood clots. But as is often the case with science, closer scrutiny of the data can alter the picture.

The bane of observational studies are confounders, variables other than the one being studied that may affect the results. Is there any reason that people who drink a glass of wine a day may be protected from cardiovascular disease other than the wine? Is it possible that they consume more fruits and vegetables and less meat as in the Mediterranean diet? Or that, like the French, they have their main meal at noon and consume fewer calories overall? Is it possible that abstainers appear to be at greater risk because they have chosen to abstain due to already having some medical condition? So, maybe the reduced early mortality in moderate alcohol consumers isn't due to alcohol but to some other factor. But that isn't even the main issue with alcohol. The issue is that the evidence is clear that alcohol is a carcinogen and that any benefit it may provide in reducing cardiovascular risk is outweighed by the risk of cancer.

In the body, ethanol is converted to acetaldehyde, the active carcinogen capable of disrupting DNA. Indeed, IARC classifies both ethanol and acetaldehyde as substances known to cause cancer in humans. But IARC just determines whether a substance is capable of causing cancer without taking extent of exposure into account! That raises the question of whether there is a threshold effect for alcohol's carcinogenicity. In other words, is there an amount of alcohol that can be ingested without a risk of triggering cancer?

The World Health Organization has amassed a significant amount of epidemiological data that seems to indicate that no amount of alcohol can be considered to be safe. Even "light" consumption, which is defined as less than 1.5 liters of wine, 3.5 liters of beer, or half a liter of spirits a week, has been linked with an increased risk of cancers of the esophagus, liver, colon, and breast. How large is this risk? Based on numerous epidemiological studies, WHO estimates that the lifetime risk of cancer for people who have one drink a day is roughly one in a hundred. That is not insignificant considering that millions of people fall into the "one-a-day" category. Even if alcohol intake were to slightly reduce the risk of heart disease, the overall impact of alcohol's effect on health is negative, especially given that in addition to cancer it increases the risk of hepatitis, fatty liver disease, pancreatitis, acid reflux, obesity, and osteoporosis.

If all that isn't enough of a deterrent, some recent studies also link alcohol consumption with accelerated aging. The DNA in our cells is the master molecule of life, controlling the production of all the proteins that are critical in biochemical processes. As we age, some of the genes in our DNA may be "switched off" and stop producing important proteins. This can happen through epigenetic changes that do not alter the basic structure of DNA but modify it by attaching to it small fragments called methyl groups. Such a modification can prevent a gene from being expressed. Blood DNA methylation can be profiled via laboratory tests, and researchers have shown that in young adults, methylation, and consequently aging, is accelerated by drinking alcohol, particularly binge drinking. If that still isn't worrisome enough, brain imaging data has also shown that alcohol can cause brain shrinkage. It isn't clear, though, what that means since Einstein's brain volume was at the lower end of normal and he did pretty well with it.

Alcohol presents other risks as well. An average of thirty-seven people die every day on American roads from alcohol related accidents. There is good evidence that the 0.08 percent alcohol limit in blood for intoxication is too high because reaction time and judgment are impaired even at this level. Then there are also deaths caused by alcohol related violence, and those linked to communicable diseases. Consider that alcohol consumption has been shown to raise the risk of HIV transmission because it increases the frequency of unprotected sex.

Does this mean that we should all become teetotalers? In terms of health, yes. But obsessing about every risk in life is also risky. Stress causes illness and enjoyable activities alleviate stress. If you enjoy that glass of wine with dinner, go for it! Recognize, though, that there is an associated risk. So, I wouldn't go for two.

And a final thought. Given the extensive attention the media has given to the contention that "no amount of alcohol is safe," I was surprised how few in my class of undergraduates knew that alcohol is a carcinogen. I encountered the same scenario at a public lecture. They just didn't know. But now, you do.

LIFESTYLE, HEART DISEASE, AND THE SWAMI

"Become a vegetarian!" was Swami Satchidananda's advice to young Dean Ornish who asked about improving his health. In 1972, Ornish was stricken with mononucleosis followed by a bout with depression that forced him to drop out of Rice University. While at home he met the swami who had been teaching meditation techniques to his sister. That sparked Dean's interest and prompted a discussion that resulted in the advice not only to embrace vegetarianism but to explore yoga and meditation as well. Ornish did, and soon felt well enough to continue his education, this time at the Baylor College of Medicine.

Seeing patients who were suffering from heart disease, often accompanied by depression, Ornish began to wonder if the program that had solved his problems would work for others. He enlisted a small group of patients who were willing to try his regimen and taught them about diet, exercise, and yoga. Within weeks they felt better, experienced less chest pain, and had lower cholesterol. This turned out to be a pivotal moment in Ornish's life and one that would define his career. He would dedicate himself to studying whether heart disease can be prevented and even reversed by diet and other lifestyle factors.

Of course, the idea that food and health are intimately connected was not novel. As early as the second century AD, the Greek physician Galen described the relationship between food and health in his work *On the Power of Foods*. Neither was the connection between diet and heart disease new. In 1908, Russian physician Alexander Ignatowski published a paper in which he described the arteries of rabbits fed a diet of full-fat milk, eggs, and meat being blocked by a buildup of fats and cholesterol. Then in 1913, Nikolai Anichkov, another Russian, showed that feeding cholesterol to rabbits causes atherosclerosis, the hardening of arteries due to a buildup of fatty deposits known as *plaque*. The question of whether dietary fats and cholesterol can also affect human arteries was not raised until the 1940s.

Dr. Lester Morrison graduated from McGill University, earned a medical degree from Temple University in Philadelphia, and went on to practice medicine in Los Angeles. Reading about Anichkov's experiments sparked an interest in heart disease that was furthered by noting records that showed a drop in the disease during wartime when food supplies became critical but increased in peacetime when food was plentiful. Since he had plenty of heart disease patients in his practice, Dr. Morrison decided to carry out an experiment. In 1946, he assigned eighty-six men and fourteen women who had suffered a heart attack to a low-cholesterol, low-fat diet and a similar number of patients who would serve as controls were asked to follow their normal diet.

Although this was not a double-blind, properly randomized trial, it was the first experimental test of the relationship between diet and heart disease in humans. After three years, body weight and blood cholesterol had dropped in the test group, anginal symptoms decreased while exercise tolerance increased. More significantly, 30 percent of the subjects in the control group died and only 14 percent in the test group. Morrison followed both groups for another five years after which the mortality rate was 44 percent in the test group and 76 percent in the control group. Judging that the study was too small and of questionable methodology, the medical establishment did not pay much attention.

In the meantime, researchers followed up on Anichkov's rabbit feeding studies. Dr. John Gofman at the University of California centrifuged the blood serum of cholesterol-fed rabbits and identified two types of cholesterol, both of which were attached to proteins that enabled them to be transported through the blood. One type floated to the top of the serum sample and was termed low-density lipoprotein cholesterol (LDL) while the other deposited at the bottom, earning the name high-density lipoprotein cholesterol (HDL). The two types of cholesterol were also found in human blood samples, and Gofman showed that men who had suffered a heart attack had elevated LDL and low HDL. Then in 1952 Dr. Laurence Kinsell published a study in the *American Journal of Clinical Nutrition* documenting a decrease in LDL with ingestion of plant foods and the avoidance of animal fats.

At this point the story took an interesting twist thanks to Nathan Pritikin, who was neither a physician nor a nutritionist. Pritikin had made a fortune as an entrepreneur and was totally shocked when in 1955, at age forty, a routine electrocardiogram revealed heart disease. The opinion at the time was that there was nothing to do except avoid both physical and mental stress. Pritikin decided to "do his own research," something we are often advised to do by today's self-appointed nutritional gurus. He came upon Dr. Morrison's small study and, since he lived nearby, made an appointment. A blood test found his cholesterol to be high, and Morrison suggested he try the low-fat diet. Pritikin went all in, adopted a strict vegetarian regimen, and, ignoring the recommendation about avoiding exertion, began a program of exercise. His blood cholesterol dropped dramatically, and he became convinced he was beating his heart disease.

A subsequent cardiac stress test during which he ran eight miles on a treadmill indeed showed normal heart function.

Thoroughly taken by his results, Pritikin became an evangelist for his diet and exercise regimen and began to fund medical studies. He educated himself thoroughly and gave talks about his experience, eventually garnering national attention, especially after the publication of his book, *The Pritikin Program for Diet and Exercise*, in 1979. Pritikin was even invited to speak at medical conferences, and although some criticized his diet as being too extreme, he earned a solid reputation in the medical community. Nathan Pritikin ended up punctuating the benefits of his program posthumously. In 1985, in the final clutches of leukemia he died of suicide, and, as required by law, an autopsy was performed. There was a near absence of atherosclerosis! Furthermore, his heart's pumping function was completely uncompromised. But a proper, randomized trial to document the reversal of heart disease by diet and exercise was still missing.

It was such a trial that Dr. Dean Ornish eventually carried out. Just a year after Pritikin's death, forty-eight patients with coronary disease were randomized into an experimental group that would follow a strict vegetarian diet, a program of exercise, as well as stress management training that involved yoga and meditation. After five years, the twenty-eight patients who followed the program had fewer cardiac events and showed regression in the blockage of coronary arteries! Although the study was small, it demonstrated that lifestyle factors can at least in some cases play an important role in reversing heart disease.

Dr. Ornish always recognized Swami Satchidananda as having inspired his research and even wrote the foreword to the swami's *Healthy Vegetarian* cookbook. Vegetarianism and yoga worked for the swami, who was healthy up to his death at age eighty-eight. Unfortunately, it seems he may have been into more than vegetarianism. His reputation is marred by allegations that he used his spiritual authority to coerce women into sexual relationships.

A CONCLUSION — KEEPING IN MIND THAT A CONCLUSION IN SCIENCE IS RARELY FINAL

Your grandmother, if you were lucky enough to have one, probably told you to eat your fruits and veggies. And now it seems that those grandmothers who meddled with our dietary habits and urged kids to eat their peas and carrots were bang on. Grandmas have been joined by a plethora of scientists who tell us that we should be eating anywhere between five and ten servings of fruits and vegetables a day. Grandmothers went by instinct, but science progresses through studies. So, what evidence do the scientists have for providing their advice?

The gold standard is the double-blind, placebo-controlled, randomized trial. Suppose you want to determine if creatine supplements aid in athletic performance. There is theoretical rationale for this since creatine occurs naturally in the body and is known to play a role in regenerating adenosine triphosphate (ATP), the molecule that provides energy for many cellular processes. A test for enhanced performance would require two groups of subjects with one group being given a fixed dose of creatine every day and the other receiving a placebo. Tests for time to exhaustion on an exercise bike, or ability to bench press weights, would be determined before and after taking the creatine supplement. A placebo control is important because just the belief that a supplement can enhance energy can actually improve performance. Double-blinding is also necessary because there is a subjective element to the determination of when a subject experiences exhaustion. Such studies have been carried out and indeed have shown that creatine supplementation can help performance in some athletic events.

When it comes to dietary studies, blinding is not possible because you know whether you are eating broccoli or a hamburger. However, this is not a big impediment because trials usually attempt to determine how dietary composition affects objective measures such as body weight, blood pressure, cholesterol levels, or onset of some disease. As an example, let's look at a trial carried out in New Zealand that aimed to investigate the effect of a plant-based diet on body weight and blood cholesterol. Sixty-five overweight subjects with at least one other cardiovascular risk factor were randomized into two groups. One group was instructed on ways to switch to a plant-based diet and for motivation told to watch *Forks Over Knives*, a documentary that features Cleveland Clinic physician Caldwell Esselstyn and Cornell University biochemist T. Colin Campbell. Both are staunch advocates of the health benefits of grains, fruits, and vegetables and believe that animal foods, including dairy, are detrimental to health. The control group was not given any specific dietary advice. After twelve months, the control group showed no change, but the experimental group had lost more weight and managed to reduce blood cholesterol.

Such studies are interesting but can be criticized for being of short duration and having too few subjects. But long-term dietary intervention studies are extremely difficult to carry out. A clear demonstration of the benefits of fruits and vegetables would require monitoring the health of two sets of subjects for decades, one consuming minimal amounts, and the other at least five servings a day. Other factors such as activity level, socioeconomic background, calorie intake, smoking, and use of medications would have to be controlled for. Clearly this represents an organizational as well as an economic challenge. That is why most of what we know about the benefits of fruits and vegetables is based on observational studies. These fall into two categories, case-control and cohort studies. In either case, there is no intervention by the researchers, they just observe a group of subjects and note their exposure to some factor of interest and record disease outcomes.

In a case-control study, subjects with a certain disease are compared to a group that is matched in every way except for the presence of the disease in question. A classic example is smoking. When groups of lung cancer patients were compared to healthy people, it became clear that the cancer patients were much more likely to be smokers. Cohort studies furthered this link. In these studies, groups of smokers and groups of non-smokers were recruited and were followed for years. The smokers were more likely to develop lung cancer.

Another type of cohort study involves following a large group of subjects for many years, evaluating their lifestyles usually through elaborate questionnaires and recording cases of disease. A classic example is the Nurses' Health Study that began in 1976 by recruiting over 120,000 registered nurses. For example, higher intake of red meat was associated with an increased the risk of premenopausal breast cancer. A Mediterranean type of diet with vegetables, nuts, and fish reduced the risk of heart disease and stroke. A high intake of green leafy vegetables reduced the risk of cognitive impairment, and high intakes of folate, vitamin B6, calcium, and vitamin D reduced the risk of colon cancer.

The Hisayama Study in Japan is another interesting example. Starting in 1961, roughly 1,000 healthy participants over the age of sixty in Hisayama filled out dietary questionnaires and were then followed for twenty-four years by which time close to half had developed some sort of dementia. Subjects who ate the most vegetables had a 30 percent reduced risk of dementia, including Alzheimer's disease, when compared with those who ate the least vegetables. There was no significant association with fruit intake. Numerous other case-control and cohort studies have shown an association between increased plant-based foods and reduced risk of various diseases.

Still, there are people who want no part of fruits or vegetables and champion a "carnivore diet," meaning they eat only meat. The claim, without any evidence, is that such a diet reduces inflammation as well as the risk of food intolerances. Proponents ignore studies that have linked meat consumption, particularly processed meat, to cancer, and run the risk of scurvy due to the lack of vitamin C and bowel problems due to a lack of fiber. To say nothing of the numerous phytochemicals that have antioxidant and other beneficial properties.

True, one can always argue that when it comes to fruits and vegetables, observational associations can never prove cause and effect and that only a randomized controlled trial can do that. However, we have such an overwhelming number of observational studies that show the benefits of fruit and vegetable intake that it would be a waste of energy and money to organize randomized trials. Grandmas were right. Eat those fruits and veggies. That doesn't mean meat has to be avoided, but best leave the carnivore diet to tigers and lions. They have no need of vitamin C or fiber.

INDEX